# A TIME FOR CREATION

*Liturgical Resources for Creation and the Environment*

Second Edition

Published by Church House Publishing
Church House
27 Great Smith Street
London SW1P 3AZ


First Edition published 2020
This Second Edition published 2026

First Edition Editors: Robert Atwell, Gill Ambrose, Christopher Irvine and Sue Moore

Some of the material in this book is extracted from *Common Worship: Times and Seasons* (2006), *Common Worship: Daily Prayer* (2005), *Common Worship: Services and Prayers for the Church of England* (2000) and *New Patterns for Worship* (2002).

ISBN 978 0 7151 2442 0



EU GPSR Authorised Representative
LOGOS EUROPE, 9 rue Nicolas Poussin, 17000, LA ROCHELLE, France
E-mail: Contact@logoseurope.eu



Cover design: www.penguinboy.net

Designed and typeset by Hugh Hillyard-Parker, Edinburgh

# A TIME FOR CREATION

*Liturgical Resources for Creation and the Environment*

Second Edition

CHURCH HOUSE PUBLISHING

# A Time for Creation

*Liturgical Resources for Creation and the Environment*

## CONTENTS

Note: The page numbers in this Second Edition match those of the original version between pages 10 and 95.

## Authorization

The material for the Festival of Creation in Christ is authorized pursuant to Canon B 2 of the Canons of the Church of England for use until further resolution of the General Synod.

The orders for Morning and Evening Prayer, Prayer During the Day, and Night Prayer comply with the provisions of A Service of the Word, which is authorized pursuant to Canon B 2 of the Canons of the Church of England for use until further resolution of the General Synod.

The order for the Eucharist complies with the provisions of The Order for the Celebration of Holy Communion also called The Eucharist and The Lord's Supper, which is authorized pursuant to Canon B 2 of the Canons of the Church of England for use until further resolution of the General Synod.

The remaining material in *A Time for Creation* (including the section 'Seasons and Festivals of the Agricultural Year' drawn from *Common Worship: Times and Seasons*, and other material drawn from elsewhere in *Common Worship* and *New Patterns for Worship*) has been commended by the House of Bishops of the General Synod for use by the minister in exercise of his or her discretion under Canon B 5 of the Canons of the Church of England.

## Copyright Information

# INTRODUCTION TO THE SECOND EDITION

> *'And he showed me a little thing, the size of a hazelnut, on the palm of my hand, round like a ball. I looked at it thoughtfully and wondered, 'What is this?' And the answer came, 'It is all that is made.' I marvelled that it continued to exist and did not suddenly disintegrate; it was so small. And again my mind supplied the answer, 'It exists, both now and for ever, because God loves it.' In short, everything owes its existence to the love of God. In this 'little thing' I saw three truths. The first is that God made it; the second is that God loves it; and the third is that God sustains it.'*[1]

In one of the most famous passages of *Revelations of Divine Love*, the medieval English writer, Julian of Norwich, distils the heart of Christian belief about the all-encompassing love of God our creator. Her words echo the theme of God's loving sovereignty proclaimed in Scripture: 'The earth is the Lord's and all that is in it, the compass of the world and all who live therein' (Psalm 24.1). They constitute a foundational truth for all Christians. We believe that God has created the world and sustains it in being. We also believe that God has uniquely entrusted the care of creation to human beings (Genesis 1.26–30) and that to God we must render an account of our stewardship. And not simply to God, but to our children and grandchildren, who are increasingly clamouring to know what will be their global inheritance.

As members of the global Anglican Communion, we are aware of those around the world who face losing their homes and livelihoods as a result of the effects of climate change. Across the world, people of different nationalities and faiths, and those who profess no faith at all, are using the language of climate emergency and environmental crisis to express their profound concern about the impact of humankind's wilful indifference to the state of the planet. Pictures of oceans clogged with discarded plastic, the desertification of productive farmland, extreme weather conditions, rising sea levels and the catastrophic decline of

[1] Julian of Norwich, *Revelations of Divine Love*, 5.

biodiversity in some parts of the world have introduced an urgency in public and private discourse.

For too long humankind has taken the environment for granted. Now that we see it threatened, we are at last waking up to the challenge of caring better for God's earth. 'To strive to safeguard the integrity of creation, and sustain and renew the life of the earth,' is one of the Five Marks of Mission.[2] If it is not embraced wholeheartedly and with determination, the other four marks of mission lack credibility. The care of creation is integral to our evangelism and mission.

## Praying with creation

The contemplation of the universe should lead us not only to the adoration of our creator, but also to take better responsibility for our actions and repent of our misuse of natural resources. Sadly, as St Augustine observes, in our self-obsession we often fail to make the connection:

> Is there anyone who, contemplating the works of God by which the entire universe is governed and ordered, is not amazed and overwhelmed by a sense of the miraculous? The power and strength of a single grain of seed is itself an amazing thing, inspiring awe in its contemplation. But humanity, preoccupied with its own petty agenda, has lost the capacity to contemplate the works of God by which it should daily render praise to God as creator.[3]

This volume of liturgical resources endeavours to help us re-forge these connections and enliven our praise of God's gifts to us in creation. It is designed to provide the parishes, schools and chaplaincies of the Church

---

[2] The five marks of mission have been developed by the Anglican Consultative Council since 1984. They have been widely adopted as an understanding of contemporary mission and were adopted by the General Synod of the Church of England in 1996. In 2012, the ACC added wording to the fourth mark, to include the need for Christians to challenge violence and work for peace.

- To proclaim the Good News of the Kingdom
- To teach, baptise and nurture new believers
- To respond to human need by loving service
- To seek to transform unjust structures of society, to challenge violence of every kind and to pursue peace and reconciliation
- To strive to safeguard the integrity of creation, and sustain and renew the life of the earth.

[3] Augustine, *Commentary on St John's Gospel, 8, 1.*

of England with a rich selection of liturgical resources with which to worship and pray, mindful of the fact that, as St Paul teaches, creation itself 'is groaning in travail' (Romans 8.22). Our prayer needs to attend to the voice of creation itself. The sense of the whole creation praying to God finds expression in the spirituality of many early Christian teachers. Tertullian, for example, in his treatise *On Prayer*, says,

> All creation prays. Cattle and wild beasts pray, and bend their knees. As they come from their barns and caves they invariably look up to heaven and call out, lifting up their spirit in their own fashion. The birds too rise and lift themselves up to heaven: instead of hands, they open out their wings in the form of a cross, and give voice to what seems to be a prayer.[4]

Gregory of Nazianzus, writing in the fourth century, says,

> All creatures praise you,
> both those who speak and those that are dumb.
> All creatures bow down before you,
> both those that can think and those that cannot.
> The longing of the universe,
> the groaning of creation reaches out to you.
> Everything that exists prays to you,
> and every creature that can read your universe
> directs to you a hymn of silence.[5]

The idea of creation reaching out to God in prayer is reflected in the liturgy of the Orthodox Church. For example, the Office Hymn for the Sunday before Lent has this verse:

> O Paradise,
> share in the sorrow of Adam who is brought to poverty,
> and with the sound of your leaves pray to the Creator
> that we may not find your gates closed for ever.
> We are fallen;
> in your compassion, have mercy on us.

Francis of Assisi also had a keen sense of creation voicing its praise to our Creator, and the repeated phrase 'praised be you' in his *Canticle of Creation* suggests that each element of creation finds a voice to praise

---

[4] Tertullian, *On Prayer*, 29.

[5] Gregory of Nazianzus, *Dogmatic Poems*, 29.

God: "Praised be you, my Lord, by our sister, mother earth, who sustains and governs us and produces various fruits and coloured flowers and grasses."[6]

## The Festival of Creation in Christ

When *A Time for Creation* was published in 2020, it offered an affirmation that the praise of God as Creator belongs at the very heart of Christian worship. The resources therein encouraged us to praise God for creation, to take responsibility for human actions in the created world, to repent of the misuse of natural resources, and to hear the voice of creation itself in prayer. Many churches had for some years celebrated a so-called 'Season of Creation' which in the Church of England did not enjoy formal status.

This Second Edition appears at a moment when these aspirations have acquired a new liturgical focus and a better-defined theological resonance. The Church of England, in companionship with many other churches, will now be celebrating the Festival of Creation in Christ instead. This Festival does not replace the many ways in which Christians have prayed, preached and acted in response to the gift and fragility of the created order. It gathers them, strengthens them, and sets them within the great confession of the Church: that all things are created through Christ, reconciled in Christ, and destined in Christ for the liberty and glory of the new creation.

The new Festival is also an ecumenical gift. The desire for a worldwide Christian feast of this kind has emerged through significant conversation amongst churches, drawing especially on the ancient and continuing witness of the Christian East and on the growing common desire of Western churches to honour God as Creator more intentionally in their calendars. The Church of England's reception of the Festival belongs within that wider Christian discernment. It is a sign that the doctrine of creation can unite Christians in praise even as it summons them together to renewed responsibility for the life of the world.

The need for such prayer is plain. We live at a time when the beauty, abundance and interdependence of the world are both more visible to us and more threatened by us. But a Christian understanding of how to

[6] Christopher Irvine, *The Cross and Creation in Christian Liturgy and Art* (Collegeville, MN: Pueblo Books, 2013), p. 202.

react to the challenges of our age must begin with praise for what God has done in creation, and in the new creation in Christ. The new Festival invites the Church to recover the doctrine of creation as a living source of worship. Creation is the gracious act by which God gives being to all things, the continuing generosity by which God sustains all things, and the theatre in which the mystery of Christ is revealed. The Father is maker of heaven and earth; the eternal Word is the one through whom all things were made; the Holy Spirit is the giver of life. When the Church celebrates creation in Christ, it is drawn into the praise of the Holy Trinity.

In the risen Christ the new creation has already begun: our broken world is being renewed, and creation itself waits with eager longing for the glorious liberty of the children of God. The Church looks back to the beginning, beholds the world held together in Christ, and looks forward to the renewal of all things in him. Without that last intention, care for creation can become anxious, burdensome or moralistic. With it, such care becomes an act of hope, a sign of reconciliation, and a participation in the renewing work of God.

This revised edition of *A Time for Creation* includes the liturgical materials for the Festival of Creation in Christ, together with commentary and supporting resources intended to help parishes, schools, chaplaincies and communities use them with confidence and imagination. These resources allow congregations to confess the ways in which we have claimed God's world as our own, to hear again that all things were created through Christ and for Christ, and to receive the promise that God makes all things new.

The Rt Revd Robert Atwell
Chair of the Liturgical Commission,
2014–2023

The Rt Revd Dr Michael Ipgrave
Chair of the Liturgical Commission,
2023–2026

# SERVICES OF THE WORD

## Morning Prayer

### Preparation

O Lord, open our lips.

*All* **And our mouth shall proclaim your praise.**

My heart tells of your word, 'Seek my face.'

*All* **Your face, Lord, will I seek.**

*Psalm 27.10*

*One of the following is said:*

1 Bless the Lord all you works of the Lord:
**Sing God's praise and exalt him for ever.**

2 Bless the Lord you heavens:
**Sing God's praise and exalt him for ever.**

3 Bless the Lord you angels of the Lord:
bless the Lord all you his hosts;
bless the Lord you waters above the heavens:
**Sing God's praise and exalt him for ever.**

4 Bless the Lord sun and moon:
bless the Lord you stars of heaven;
bless the Lord all rain and dew:
**Sing God's praise and exalt him for ever.**

5 Bless the Lord all winds that blow:
bless the Lord you fire and heat;
bless the Lord scorching wind and bitter cold:
**Sing God's praise and exalt him for ever.**

6 Bless the Lord dews and falling snows:
bless the Lord you nights and days;
bless the Lord light and darkness:
**Sing God's praise and exalt him for ever.**

7 Bless the Lord frost and cold:
bless the Lord you ice and snow;
bless the Lord lightnings and clouds:
**Sing God's praise and exalt him for ever.**

8 O let the earth bless the Lord:
bless the Lord you mountains and hills;
bless the Lord all that grows in the ground:
**Sing God's praise and exalt him for ever.**

9 Bless the Lord you springs:
bless the Lord you seas and rivers;
bless the Lord you whales and all that swim in the waters:
**Sing God's praise and exalt him for ever.**

10 Bless the Lord all birds of the air:
bless the Lord you beasts and cattle;
bless the Lord all people on earth:
**Sing God's praise and exalt him for ever.**

11 O people of God bless the Lord:
bless the Lord you priests of the Lord;
bless the Lord you servants of the Lord:
**Sing God's praise and exalt him for ever.**

12 Bless the Lord all you of upright spirit:
bless the Lord you that are holy and humble in heart.

*The Song of the Three 35–65*

*All* **Bless the Father, the Son and the Holy Spirit:**
**Sing God's praise and exalt him for ever.**

*(or)*

Blessed are you, Lord God,
creator of heaven and earth.
Your Word calls all things into being,
and the light of dawn awakens us to life.
May your wisdom guide us this day
that we may cherish and care for your good creation,
and offer to you the sacrifice of our lips,
praising you, Father, Son and Holy Spirit.
*All* **Blessed be God for ever.**

The night has passed, and the day lies open before us;
let us pray with one heart and mind.

*Silence is kept.*

As we rejoice in the gift of this new day,
so may the light of your presence, O God,
set our hearts on fire with love for you;
now and for ever.

*All* **Amen.**

## The Word of God

### *Psalmody*

*One of the psalms suggested on page 34 or Psalm 148*

1 Alleluia.
Praise the Lord from the heavens; ♦
praise him in the heights.

2 Praise him, all you his angels; ♦
praise him, all his host.

3 Praise him, sun and moon; ♦
praise him, all you stars of light.

4 Praise him, heaven of heavens, ♦
and you waters above the heavens.

5 Let them praise the name of the Lord, ♦
for he commanded and they were created.

6 He made them fast for ever and ever; ♦
he gave them a law which shall not pass away.

7 Praise the Lord from the earth, ♦
you sea monsters and all deeps;

8 Fire and hail, snow and mist, ♦
tempestuous wind, fulfilling his word;

9 Mountains and all hills, ♦
fruit trees and all cedars;

10 Wild beasts and all cattle, ♦
creeping things and birds on the wing;

11 Kings of the earth and all peoples, ♦
princes and all rulers of the world;

12 Young men and women,
old and young together; ♦
let them praise the name of the Lord.

13 For his name only is exalted, ♦
his splendour above earth and heaven.

14 He has raised up the horn of his people
and praise for all his faithful servants, ♦
the children of Israel, a people who are near him.
Alleluia.

*Glorious God,*
*your whole creation sings your marvellous work;*
*may heaven's praise so echo in our hearts*
*that we may be good stewards of the earth;*
*through Jesus Christ our Lord.*

*If the psalm prayer is not used, the psalm ends with*

***All*** **Glory to the Father and to the Son**
**and to the Holy Spirit;**
**as it was in the beginning is now**
**and shall be for ever. Amen.**

## *Scripture Reading*

*One (or two) of the readings suggested on pages 120–121*
*or the following*

In the beginning when God created the heavens and the earth, the earth was a formless void and darkness covered the face of the deep, while a wind from God swept over the face of the waters. Then God said, 'Let there be light'; and there was light. And God saw that the light was good; and God separated the light from the darkness. God called the light Day, and the darkness he called Night. And there was evening and there was morning, the first day.

*Genesis 1.1–5*

*One of the Canticles on pages 35–41 may also be said or sung before or between the readings.*

*The reading(s) may be followed by a time of silence.*

*A suitable song or chant, or a responsory in this or another form, may follow*

The heavens proclaim the glory of God:
and the firmament shows his handiwork.

*All* **The firmament shows God's handiwork:**
**And one day pours out its song to another.**

The sun goes forth from the ends of the heavens.

*All* **and proclaims the glory of God.**

Glory to the Father and to the Son and to the Holy Spirit.

*All* **The heavens proclaim the glory of God:**
**and the firmament shows his handiwork.**

*Psalm 19*

## *Gospel Canticle*

*The Benedictus (The Song of Zechariah) is said.*

**Let the fields be joyful and all that is in them:**
**let all the trees of the wood shout for joy before the Lord.**

1 Blessed be the Lord the God of Israel, ♦
who has come to his people and set them free.

2 He has raised up for us a mighty Saviour, ♦
born of the house of his servant David.

3 Through his holy prophets God promised of old ♦
to save us from our enemies,
from the hands of all that hate us,

4 To show mercy to our ancestors, ♦
and to remember his holy covenant.

5 This was the oath God swore to our father Abraham: ♦
to set us free from the hands of our enemies,

6 Free to worship him without fear, ♦
holy and righteous in his sight
all the days of our life.

7 And you, child, shall be called the prophet of the Most High, ♦
for you will go before the Lord to prepare his way,

8 To give his people knowledge of salvation ♦
by the forgiveness of all their sins.

9 In the tender compassion of our God ♦
the dawn from on high shall break upon us,

10 To shine on those who dwell in darkness and the shadow of death, ♦
and to guide our feet into the way of peace.

*Luke 1.68–79*

*All* **Glory to the Father and to the Son**
**and to the Holy Spirit;**
**as it was in the beginning is now**
**and shall be for ever. Amen.**

**Let the fields be joyful and all that is in them:**
**let all the trees of the wood shout for joy before the Lord.**

## Prayers

*One of the forms on pages 65–72 or other suitable prayers may be used.*

*This response may be used*

God of life:
*All* **hear our prayer.**

*The Collect is said*

Lord our God,
as with all creation
we offer you the life of this new day,
give us grace to love and serve you
to the praise of Jesus Christ our Lord.
*All* **Amen.**

*The Lord's Prayer is said*

As our Saviour taught us, so we pray

*All* **Our Father in heaven,**
**hallowed be your name,**
**your kingdom come,**
**your will be done,**
**on earth as in heaven.**
**Give us today our daily bread.**
**Forgive us our sins**
**as we forgive those who sin against us.**
**Lead us not into temptation**
**but deliver us from evil.**
**For the kingdom, the power,**
**and the glory are yours**
**now and for ever.**
**Amen.**

## The Conclusion

May God who made heaven and earth bless us and keep us.

*All* **Amen.**

Let us bless the Lord.

*All* **Thanks be to God.**

# Prayer During the Day

## Preparation

O God, make speed to save us.
*All* **O Lord, make haste to help us.**

Send forth your Spirit, O Lord
*All* **and renew the face of the earth.**

*from Psalm 104*

## Praise

*A hymn, song, canticle, extempore praise or*

We praise you, O God,
we acclaim you as the Lord;
all creation worships you,
the Father everlasting.
To you all angels, all the powers of heaven,
the cherubim and seraphim, sing in endless praise:
Holy, holy, holy Lord, God of power and might,
heaven and earth are full of your glory.

*from Te Deum Laudamus*

## The Word of God

### *Psalmody*

*One of the psalms suggested on page 34 or Psalm 19*

1 The heavens are telling the glory of God ♦
and the firmament proclaims his handiwork.

2 One day pours out its song to another ♦
and one night unfolds knowledge to another.

3 They have neither speech nor language ♦
and their voices are not heard,

4 Yet their sound has gone out into all lands ♦
and their words to the ends of the world.

5 In them has he set a tabernacle for the sun, ♦
that comes forth as a bridegroom out of his chamber
and rejoices as a champion to run his course.

6 It goes forth from the end of the heavens
and runs to the very end again, ♦
and there is nothing hidden from its heat.

7 The law of the Lord is perfect, reviving the soul; ♦
the testimony of the Lord is sure
and gives wisdom to the simple.

8 The statutes of the Lord are right and rejoice the heart; ♦
the commandment of the Lord is pure
and gives light to the eyes.

9 The fear of the Lord is clean and endures for ever; ♦
the judgements of the Lord are true
and righteous altogether.

10 More to be desired are they than gold,
more than much fine gold, ♦
sweeter also than honey,
dripping from the honeycomb.

11 By them also is your servant taught ♦
and in keeping them there is great reward.

12 Who can tell how often they offend? ♦
O cleanse me from my secret faults!

13 Keep your servant also from presumptuous sins
lest they get dominion over me; ♦
so shall I be undefiled,
and innocent of great offence.

14 Let the words of my mouth and the meditation of my heart
be acceptable in your sight, ♦
O Lord, my strength and my redeemer.

*Christ, the sun of righteousness,*
*rise in our hearts this day,*
*enfold us in the brightness of your love*
*and bear us at the last to heaven's horizon;*
*for your love's sake.*

*If the psalm prayer is not used, the psalm ends with*

***All*** **Glory to the Father and to the Son**
**and to the Holy Spirit;**
**as it was in the beginning is now**
**and shall be for ever. Amen.**

### *Short Reading*

*One of the readings suggested on pages 120–121 or*

The Lord God said, 'As long as the earth endures, seedtime and harvest, cold and heat, summer and winter, day and night, shall not cease.' Then God said to Noah and to his sons with him, 'As for me, I am establishing my covenant with you and your descendants after you, and with every living creature that is with you, the birds, the domestic animals, and every animal of the earth with you, as many as came out of the ark.

*Genesis 8.22; 9.8–10*

### *Response*

*Silence, study, song, or words from Scripture (see pages 42–43), such as*

You are worthy, our Lord and God,
***All*** **to receive glory and honour and power.**

For you created all things,
***All*** **and by your will they existed and were created.**

*Revelation 4.11*

## Prayers

*One of the forms on pages 65–72 or other suitable prayers may be used.*

*This response may be used*

God of life:
***All*** **hear our prayer.**

*One of the Collects on page 119, or the Collect of the Day, or this Collect is said*

Eternal Lord,
our beginning and our end:
bring us with the whole creation
to your glory, hidden through past ages
and made known
in Jesus Christ our Lord.

*All* **Amen.**

*The Lord's Prayer is said*

As our Saviour taught us, so we pray

*All* **Our Father in heaven,**
**hallowed be your name,**
**your kingdom come,**
**your will be done,**
**on earth as in heaven.**
**Give us today our daily bread.**
**Forgive us our sins**
**as we forgive those who sin against us.**
**Lead us not into temptation**
**but deliver us from evil.**
**For the kingdom, the power,**
**and the glory are yours**
**now and for ever.**
**Amen.**

## The Conclusion

May God who made heaven and earth bless us and keep us.

*All* **Amen.**

Let us bless the Lord.

*All* **Thanks be to God.**

# Evening Prayer

## Preparation

O God, make speed to save us.
*All* **O Lord, make haste to help us.**

My heart tells of your word, 'Seek my face.'
*All* **Your face, Lord, will I seek.**

*Psalm 27.10*

*One of the following is said*

Blessed are you, Lord God, King of the universe.
*All* **Your word brings on the dusk of evening.**

Your wisdom creates both night and day.
*All* **You determine the cycle of time.**

You arrange the succession of seasons
and establish the stars in their heavenly courses.
*All* **The Lord of hosts is your name.**

Living and eternal God, rule over us always.
*All* **Blessed be the Lord, whose word makes evening fall.**

*or*

Blessed are you, Lord of life.
You make light and darkness,
morning and evening, day and night,
and establish the stars in their heavenly courses.
As evening falls and creatures move on the face of the earth
may we take our rest,
and awake refreshed to praise you,
God, Father, Son and Holy Spirit.
*All* **Blessed be God for ever.**

That this evening may be holy, good and peaceful,
let us pray with one heart and mind.

*Silence is kept.*

As our evening prayer rises before you, O God,
so may your mercy come down upon us
to cleanse our hearts
and set us free to sing your praise
now and for ever.

*All* **Amen.**

## The Word of God

### *Psalmody*

*One of the psalms suggested on page 34 or Psalm 104.1,21–32*

1 Bless the Lord, O my soul. ♦
O Lord my God, how excellent is your greatness!

21 You appointed the moon to mark the seasons, ♦
and the sun knows the time for its setting.

22 You make darkness that it may be night, ♦
in which all the beasts of the forest creep forth.

23 The lions roar for their prey ♦
and seek their food from God.

24 The sun rises and they are gone ♦
to lay themselves down in their dens.

25 People go forth to their work ♦
and to their labour until the evening.

26 O Lord, how manifold are your works! ♦
In wisdom you have made them all;
the earth is full of your creatures.

27 There is the sea, spread far and wide, ♦
and there move creatures beyond number, both small and great.

28 There go the ships, and there is that Leviathan ♦
which you have made to play in the deep.

29 All of these look to you ♦
to give them their food in due season.

30 When you give it them, they gather it; ♦
you open your hand and they are filled with good.

31 When you hide your face they are troubled; ♦
when you take away their breath,
they die and return again to the dust.

32 When you send forth your spirit, they are created, ♦
and you renew the face of the earth.

*Creator God,*
*send your Holy Spirit to renew this living world,*
*that the whole creation,*
*in its groaning and striving,*
*may know your loving purpose*
*and come to reflect your glory;*
*in Jesus Christ our Lord. Amen.*

*If the psalm prayer is not used, the psalm ends with*

***All*** **Glory to the Father and to the Son**
**and to the Holy Spirit;**
**as it was in the beginning is now**
**and shall be for ever. Amen**

## *Scripture Reading*

*One (or two) of the readings suggested on pages 120–121 or the following*

I consider that the sufferings of this present time are not worth comparing with the glory about to be revealed to us. For the creation waits with eager longing for the revealing of the children of God; for the creation was subjected to futility, not of its own will but by the will of the one who subjected it, in hope that the creation itself will be set free from its bondage to decay and will obtain the freedom of the glory of the children of God. We know that the whole creation has been groaning in labour pains until now.

*Romans 8.18–22*

*One of the Canticles on pages 35–41 may also be said or sung before or between the readings.*

*The reading(s) may be followed by a time of silence.*

*A suitable song or chant, or a responsory in this or another form, may follow*

I will sing for ever of your love, O Lord,
*All* **my lips shall proclaim your faithfulness.**

The heavens bear witness to your wonders;
*All* **I will sing for ever of your love, O Lord.**

The assembly of your saints proclaims your truth;
*All* **my lips shall proclaim your faithfulness.**

Righteousness and justice are the foundation of your throne;
steadfast love and faithfulness go before you.
*All* **I will sing for ever of your love, O Lord,**
**my lips shall proclaim your faithfulness.**

*from Psalm 89*

## *Gospel Canticle*

*The Magnificat (The Song of Mary) is said*

**The day and the night are yours, O God,**
**and the sun, the moon and the stars reflect your glory.**

1 My soul proclaims the greatness of the Lord,
  my spirit rejoices in God my Saviour; ♦
he has looked with favour on his lowly servant.

2 From this day all generations will call me blessed; ♦
the Almighty has done great things for me
  and holy is his name.

3 He has mercy on those who fear him, ♦
from generation to generation.

4 He has shown strength with his arm ♦
and has scattered the proud in their conceit,

5 Casting down the mighty from their thrones ♦
and lifting up the lowly.

6 He has filled the hungry with good things ♦
and sent the rich away empty.

7 He has come to the aid of his servant Israel, ♦
to remember his promise of mercy,

8 The promise made to our ancestors, ♦
to Abraham and his children for ever.

*Luke 1.46-55*

*All* **Glory to the Father and to the Son**
**and to the Holy Spirit;**
**as it was in the beginning is now**
**and shall be for ever. Amen.**

**The day and the night are yours, O God,**
**and the sun, the moon and the stars reflect your glory.**

## Prayers

*One of the forms on pages 65–72 or other suitable prayers may be used.*

*This response may be used*

God of life:
*All* **hear our prayer.**

*One of the Collects on page 119 or the Collect of the Day, or this Collect is said*

Eternal Lord,
our beginning and our end:
bring us with the whole creation
to your glory, hidden through past ages
and made known
in Jesus Christ our Lord.
*All* **Amen.**

*The Lord's Prayer is said*

As our Saviour taught us, so we pray

*All* **Our Father in heaven,**
**hallowed be your name,**
**your kingdom come,**
**your will be done,**
**on earth as in heaven.**
**Give us today our daily bread.**
**Forgive us our sins**
**as we forgive those who sin against us.**
**Lead us not into temptation**
**but deliver us from evil.**
**For the kingdom, the power,**
**and the glory are yours**
**now and for ever.**
**Amen.**

## The Conclusion

May God who made heaven and earth bless us and keep us.

*All* **Amen.**

Let us bless the Lord

*All* **Thanks be to God.**

# Night Prayer (Compline)

## Preparation

The Lord almighty grant us a quiet night and a perfect end.

*All* **Amen.**

Our help is in the name of the Lord

*All* **who made heaven and earth.**

*A period of silence for reflection on the past day may follow.*

*The following or other suitable words of penitence may be used*

We confess to you our lack of care
for the world you have given us.
Lord have mercy,

*All* **Lord have mercy.**

We confess to you our selfishness
in not sharing the earth's bounty fairly.
Christ have mercy,

*All* **Christ have mercy.**

We confess to you our failure
to protect resources for others,
Lord have mercy,

*All* **Lord have mercy.**

May God who loved the world so much
that he sent his Son to be our Saviour
forgive us our sins
and make us holy to serve him in the world,
through Jesus Christ our Lord.

*All* **Amen.**

O God, make speed to save us.

*All* **O Lord, make haste to help us.**

*All* **Glory to the Father and to the Son**
**and to the Holy Spirit;**
**as it was in the beginning is now**
**and shall be for ever. Amen. [Alleluia.]**

*The following or another suitable hymn may be sung*

The spacious firmament on high,
with all the blue ethereal sky,
and spangled heavens, a shining frame,
their great Original proclaim.

The unwearied sun from day to day
does his Creator's power display,
and publishes to every land
the works of an almighty hand.

Soon as the evening shades prevail
the moon takes up the wondrous tale,
and nightly to the listening earth
repeats the story of her birth;

Whilst all the stars that round her burn,
and all the planets in their turn,
confirm the tidings, as they roll,
and spread the truth from pole to pole.

What though in solemn silence all
move round the dark terrestrial ball;
what though nor real voice nor sound
amid their radiant orbs he found;

In reason's ear they all rejoice,
and utter forth a glorious voice;
for ever singing as they shine,
'the hand that made us is divine.'

*(May be sung to Te Lucis Ante Terminum, Mode viii)*

## The Word of God

*Psalmody*

*One or both of the following psalms may be used.*

**Psalm 65**

1 Praise is due to you, O God, in Zion; ♦
to you that answer prayer shall vows be paid.

2 To you shall all flesh come to confess their sins; ♦
when our misdeeds prevail against us,
    you will purge them away.

3 Happy are they whom you choose
    and draw to your courts to dwell there. ♦
We shall be satisfied with the blessings of your house,
    even of your holy temple.

4 With wonders you will answer us in your righteousness,
    O God of our salvation, ♦
O hope of all the ends of the earth
    and of the farthest seas.

5 In your strength you set fast the mountains ♦
and are girded about with might.

6 You still the raging of the seas, ♦
the roaring of their waves
    and the clamour of the peoples.

7 Those who dwell at the ends of the earth
    tremble at your marvels; ♦
the gates of the morning and evening sing your praise.

8 You visit the earth and water it; ♦
    you make it very plenteous.

9 The river of God is full of water; ♦
you prepare grain for your people,
    for so you provide for the earth.

10 You drench the furrows and smooth out the ridges; ♦
you soften the ground with showers and bless its increase.

11 You crown the year with your goodness, ♦
and your paths overflow with plenty.

12 May the pastures of the wilderness flow with goodness ♦
and the hills be girded with joy.

13 May the meadows be clothed with flocks of sheep ♦
and the valleys stand so thick with corn
    that they shall laugh and sing.

*or*

**Psalm 134**

1 Come, bless the Lord, all you servants of the Lord, ♦
you that by night stand in the house of the Lord.

2 Lift up your hands towards the sanctuary ♦
and bless the Lord.

3 The Lord who made heaven and earth ♦
give you blessing out of Zion.

*At the end of the psalmody, the following is said or sung*

*All* **Glory to the Father and to the Son**
**and to the Holy Spirit;**
**as it was in the beginning is now**
**and shall be for ever. Amen.**

## *Scripture Reading*

*One of the following short lessons or another suitable passage is read.*

God saw everything that he had made, and indeed, it was very good.

*Genesis 1.31*

O Lord, how manifold are your works! In wisdom you have made them all; the earth is full of your creatures.

*Psalm 104.26*

In Christ all things in heaven and on earth were created, things visible and invisible ... all things have been created through him and for him.

*Colossians 1.16*

You are worthy, our Lord and God, to receive glory and honour and power, for you created all things, and by your will they existed and were created.

*Revelation 4.11*

## *Response*

*The following responsory may be said*

Into your hands, O Lord, I commend my spirit.

*All* **Into your hands, O Lord, I commend my spirit.**

For you have redeemed me, Lord God of truth.

*All* **I commend my spirit.**

Glory to the Father and to the Son
and to the Holy Spirit.

*All* **Into your hands, O Lord, I commend my spirit.**

Keep me, O Lord, as the apple of your eye. [Alleluia.]

*All* **Hide me under the shadow of your wings. [Alleluia.]**

## *The Gospel Canticle*

*The Nunc dimittis (the Song of Simeon) is said or sung*

**In you, O Christ, there is a new creation.
The old has passed away, and all things are being made anew.**

Now, Lord, you let your servant go in peace: ♦
your word has been fulfilled.

My own eyes have seen the salvation ♦
which you have prepared in the sight of every people;

A light to reveal you to the nations ♦
and the glory of your people Israel.

*Luke 2.29–32*

*All* **Glory to the Father and to the Son
and to the Holy Spirit;
as it was in the beginning is now
and shall be for ever. Amen.**

**In you, O Christ, there is a new creation.
The old has passed away, and all things are being made anew.**

*2 Corinthians 5.17 (adapted)*

## The Prayers

*Intercessions and thanksgivings may be offered here.*

*Collect*

Heavenly Father and Lord of Life,
you have created the heavens and the earth,
the sun and the earth, the moon and the stars.
Grant that as now we take our rest,
we may awake strengthened to serve you
in the care of your good creation,
and with a song of praise on our lips;
through Jesus Christ, our Lord.

*All* **Amen.**

*The Lord's Prayer may be said.*

*All* **Our Father, who art in heaven,**
**hallowed be thy name;**
**thy kingdom come;**
**thy will be done;**
**on earth as it is in heaven.**
**Give us this day our daily bread.**
**And forgive us our trespasses,**
**as we forgive those who trespass against us.**
**And lead us not into temptation;**
**but deliver us from evil.**
**For thine is the kingdom,**
**the power and the glory,**
**for ever and ever.**
**Amen.**

## *The Conclusion*

In peace we will lie down and sleep;

*All* **for you alone, Lord, make us dwell in safety.**

Abide with us, Lord Jesus,

*All* **for the night is at hand and the day is now past.**

As the night watch looks for the morning,

*All* **so do we look for you, O Christ.**

[Come with the dawning of the day

*All* **and make yourself known in the breaking of the bread.]**

The Lord bless us and watch over us;
the Lord make his face shine upon us and be gracious to us;
the Lord look kindly on us and give us peace.

*All* **Amen.**

# ALTERNATIVE PSALMS AND CANTICLES IN DAILY PRAYER

## Psalms

| | |
|---|---|
| 8 | O Lord our governor |
| 19 | The heavens are telling the glory of God |
| 24 | The earth is the Lord's |
| 33 | Rejoice in the Lord, O you righteous |
| 65 | Praise is due to you, O God, in Zion |
| 80 | Hear, O Shepherd of Israel |
| 90.1–6, 13–end | Lord, you have been our refuge |
| 96 | Sing to the Lord a new song |
| 100 | O be joyful in the Lord |
| 103 | Bless the Lord, O my soul |
| 104.25–37 | People go forth to their work |
| 106 | Alleluia. Give thanks to the Lord, for he is gracious |
| 107 | O give thanks to the Lord, for he is gracious |
| 112 | Alleluia. Blessed are those who fear the Lord |
| 126 | When the Lord restored the fortunes of Zion |
| 136.1–9, 23–26 | Give thanks to the Lord, for he is gracious |
| 146 | Alleluia. Praise the Lord, O my soul |
| 147.1–13 | Alleluia. How good it is to make music for our God |
| 148 | Alleluia. Praise the Lord from the heavens |

# Canticles

## *A Song of the Wilderness*

*All* **Lift up your voice with strength,**
**O herald of good tidings.**

1 The wilderness and the dry land shall rejoice, ♦
the desert shall blossom and burst into song.

2 They shall see the glory of the Lord, ♦
the majesty of our God.

3 Strengthen the weary hands, ♦
and make firm the feeble knees.

4 Say to the anxious, 'Be strong, fear not,
your God is coming with judgement, ♦
coming with judgement to save you.'

5 Then shall the eyes of the blind be opened, ♦
and the ears of the deaf unstopped;

6 Then shall the lame leap like a hart, ♦
and the tongue of the dumb sing for joy.

7 For waters shall break forth in the wilderness, ♦
and streams in the desert;

8 The ransomed of the Lord shall return with singing, ♦
with everlasting joy upon their heads.

9 Joy and gladness shall be theirs, ♦
and sorrow and sighing shall flee away.

*Isaiah 35.1,2b–4a,4c–6,10*

*All* **Glory to the Father and to the Son**
**and to the Holy Spirit;**
**as it was in the beginning is now**
**and shall be for ever. Amen.**

*All* **Lift up your voice with strength,**
**O herald of good tidings.**

## *A Song of the New Creation*

*All* **I will make a way in the wilderness,**
**and rivers in the desert.**

1 'I am the Lord, your Holy One, ♦
the Creator of Israel, your King.'

2 Thus says the Lord, who makes a way in the sea, ♦
a path in the mighty waters,

3 'Remember not the former things, ♦
nor consider the things of old.

4 'Behold, I am doing a new thing; ♦
now it springs forth, do you not perceive it?

5 'I will make a way in the wilderness
and rivers in the desert, ♦
to give drink to my chosen people,

6 'The people whom I formed for myself, ♦
that they might declare my praise.'

*Isaiah 43.15,16,18,19,20c,21*

*All* **Glory to the Father and to the Son**
**and to the Holy Spirit;**
**as it was in the beginning is now**
**and shall be for ever. Amen.**

*All* **I will make a way in the wilderness,**
**and rivers in the desert.**

## *A Song of Creation (Benedicite) (a)*

1 Bless the Lord all you works of the Lord:
Sing God's praise and exalt him for ever.

2 Bless the Lord you heavens:
Sing God's praise and exalt him for ever.

3 Bless the Lord you angels of the Lord:
bless the Lord all you his hosts;
bless the Lord you waters above the heavens:
Sing God's praise and exalt him for ever.

4 Bless the Lord sun and moon:
bless the Lord you stars of heaven;
bless the Lord all rain and dew:
Sing God's praise and exalt him for ever.

5 Bless the Lord all winds that blow:
bless the Lord you fire and heat;
bless the Lord scorching wind and bitter cold:
Sing God's praise and exalt him for ever.

6 Bless the Lord dews and falling snows:
bless the Lord you nights and days;
bless the Lord light and darkness:
Sing God's praise and exalt him for ever.

7 Bless the Lord frost and cold:
bless the Lord you ice and snow;
bless the Lord lightnings and clouds:
Sing God's praise and exalt him for ever.

8 O let the earth bless the Lord:
bless the Lord you mountains and hills;
bless the Lord all that grows in the ground:
Sing God's praise and exalt him for ever.

9 Bless the Lord you springs:
bless the Lord you seas and rivers;
bless the Lord you whales and all that swim in the waters:
Sing God's praise and exalt him for ever.

10 Bless the Lord all birds of the air:
bless the Lord you beasts and cattle;
bless the Lord all people on earth:
Sing God's praise and exalt him for ever.

11 O people of God bless the Lord:
bless the Lord you priests of the Lord;
bless the Lord you servants of the Lord:
Sing God's praise and exalt him for ever.

12 Bless the Lord all you of upright spirit:
bless the Lord you that are holy and humble in heart.

*The Song of the Three 35–65*

*All* **Bless the Father, the Son and the Holy Spirit:**
**Sing God's praise and exalt him for ever.**

## *A Song of Creation (Benedicite) (b)*

1 Bless the Lord all you works of the Lord: ♦
sing his praise and exalt him for ever.

2 Bless the Lord you heavens: ♦
sing his praise and exalt him for ever.

3 Bless the Lord you angels of the Lord: ♦
sing his praise and exalt him for ever.

4 Bless the Lord all people on earth: ♦
sing his praise and exalt him for ever.

5 O people of God bless the Lord: ♦
sing his praise and exalt him for ever.

6 Bless the Lord you priests of the Lord: ♦
sing his praise and exalt him for ever.

7 Bless the Lord you servants of the Lord: ♦
sing his praise and exalt him for ever.

8 Bless the Lord all you of upright spirit: ♦
bless the Lord you that are holy and humble in heart.

*The Song of the Three 35–37, 60–65*

*All* **Bless the Father, the Son and the Holy Spirit: ♦**
**sing his praise and exalt him for ever.**

## *A Song of Praise*

*All* **You created all things, O God,**
**and are worthy of our praise for ever.**

1 You are worthy, our Lord and God, ♦
to receive glory and honour and power.

2 For you have created all things, ♦
and by your will they have their being.

3 You are worthy, O Lamb, for you were slain, ♦
and by your blood you ransomed for God
saints from every tribe and language and nation.

4 You have made them to be a kingdom and priests
serving our God, ♦
and they will reign with you on earth.

*Revelation 4.11; 5.9b,10*

*All* **To the One who sits on the throne and to the Lamb ♦**
**be blessing and honour and glory and might,**
**for ever and ever. Amen.**

*All* **You created all things, O God,**
**and are worthy of our praise for ever.**

## *A Song of the Holy City*

*All* **I saw the holy city**
**coming down out of heaven from God.**

1 I saw a new heaven and a new earth, ♦
for the first heaven and the first earth had passed away
and the sea was no more.

2 And I saw the holy city, new Jerusalem,
coming down out of heaven from God, ♦
prepared as a bride adorned for her husband.

3 And I heard a great voice from the throne saying, ♦
'Behold, the dwelling of God is among mortals.

4 'He will dwell with them and they shall be his peoples, ♦
and God himself will be with them.

5 'He will wipe every tear from their eyes, ♦
and death shall be no more.

6 'Neither shall there be mourning,
nor crying, nor pain any more, ♦
for the former things have passed away.'

7 And the One who sat upon the throne said, ♦
'Behold, I make all things new.'

*Revelation 21.1–5a*

*All* **To the One who sits on the throne and to the Lamb ♦**
**be blessing and honour and glory and might,**
**for ever and ever. Amen.**

*All* **I saw the holy city**
**coming down out of heaven from God.**

## *A Song of Francis of Assisi*

1 Most High, all powerful, good Lord, ♦
to you be praise, glory, honour and all blessing.

2 Only to you, Most High, do they belong ♦
and no one is worthy to call upon your name.

3 May you be praised, my Lord, with all your creatures,
especially brother sun, ♦
through whom you lighten the day for us.

4 He is beautiful and radiant with great splendour; ♦
he signifies you, O Most High.

5 Be praised, my Lord, for sister moon and the stars: ♦
clear and precious and lovely, they are formed in heaven.

6 Be praised, my Lord, for brother wind; ♦
for air and clouds, clear skies and all weathers,
by which you give sustenance to your creatures.

7 Be praised, my Lord, for sister water, ♦
who is very useful and humble and precious and pure.

8 Be praised, my Lord, for brother fire,
by whom the night is illumined for us: ♦
he is beautiful and cheerful, full of power and strength.

9 Be praised, my Lord, for our sister, mother earth,
who sustains and governs us ♦
and produces diverse fruits
and coloured flowers and grass.

10 Be praised, my Lord, by all those who forgive for love of you ♦
and who bear weakness and tribulation.

11 Blessed are those who bear them in peace: ♦
for by you, Most High, they will be crowned.

12 Be praised, my Lord, for our sister, the death of the body,
from which no one living is able to flee; ♦
woe to those who are dying in mortal sin.

13 Blessed are those who are found doing your most holy will, ♦
for the second death will do them no harm.

14 Praise and bless my Lord and give him thanks ♦
and serve him with great humility.

# SHORT PASSAGES OF SCRIPTURE

1 God saw everything that he had made,
and indeed, it was very good. *Genesis 1.31*

2 Let the peoples praise you, O God;
let all the peoples praise you.
The earth has yielded its increase;
God, our God, has blessed us. *Psalm 67.5,6*

3 The Lord will indeed give all that is good,
and our land will yield its increase. *Psalm 85.12*

4 O Lord, how manifold are your works!
In wisdom you have made them all;
the earth is full of your creatures. *Psalm 104.26*

5 Those who are generous are blessed,
for they share their bread with the poor. *Proverbs 22.9*

6 For as the rain and the snow come down from heaven,
and do not return there until they have watered the earth,
making it bring forth and sprout,
giving seed to the sower and bread to the eater,
so shall my word be that goes out from my mouth;
it shall not return to me empty,
but it shall accomplish that which I purpose,
and succeed in the thing for which I sent it.
*Isaiah 55.10–11*

7 Sow for yourselves righteousness;
reap steadfast love;
break up your fallow ground;
for it is time to seek the Lord,
that he may come and rain righteousness upon you. *Hosea 10.12*

8 He who supplies seed to the sower and bread for food
will supply and multiply your seed for sowing
and increase the harvest of your righteousness.
*2 Corinthians 9.10*

9 In Christ all things in heaven and on earth were created,
things visible and invisible ... all things have been created
through him and for him.

*Colossians 1.16*

10 You are worthy, our Lord and God,
to receive glory and honour and power,
for you created all things,
and by your will they existed and were created. *Revelation 4.11*

# PRAYERS FOR THE ENVIRONMENT

## *A lament for a time of global environmental and climate crisis*

The following lament has been composed from various scriptural texts to voice our grief in the face of this global environmental crisis. 'Take words with you and return to the Lord; say to him, "Take away all guilt; accept that which is good, and we will offer the fruit of our lips" (Hosea 14.2). As the Psalms give witness, in worship we come before the Lord not only with words of praise, but also of lament to express our grief and sorrow.

In shaping its use in worship, it may be appropriate for the leader to keep a space for silence either between the verses, for reflection, or before the lament is recited. Such a period of silence may conclude with the words, 'Lord, hear our cry'.

***Refrain***

*All* **The land cries out to you, O Lord.**

1 The land mourns: and all who dwell on it languish.

2 The beasts of the field, the birds of the air: and even the fish of the sea are perishing. *(Hosea 4.3)*

3 For the earth is scorched with fire: the air polluted, and the waters choked with waste.

4 The ice melts and the seas rage: the waters surge and floods devastate the land.

5 The water springs are silent, and the steams run dry: the green places are a dry land where there is no water. *(Psalm 63.2b)*

6 The fertile ground is exhausted: the land has become a parched and windswept wasteland.

7 The holy cities have become a wilderness: Zion has become a desolation. *(Isaiah 64.10)*

8 So turn us again, O Lord: and come to our help. *(Psalm 80.4a,3b)*

9 Drop down, O heavens, from above: and let righteousness rain down upon the poor. *(Isaiah 45.8)*

10 Let the earth open: and let wholeness and healing spring forth. *(Isaiah 45.8b)*

11 Then the wilderness and the dry land shall be glad:
the desert shall rejoice and blossom. *(Isaiah 35.1)*

12 The mountains and hills shall break forth into singing:
and the trees of the forest shall clap their hands. *(Isaiah 55.12b)*

*All* **Glory to the Father and to the Son**
**and to the Holy Spirit;**
**as it was in the beginning is now**
**and shall be for ever. Amen.**

*All* **The land cries out to you, O Lord.**

## *A prayer for the farming community in a time of crisis*

God of life and hope,
you sent us from the garden of Eden
to till the soil and to grow our food by the sweat of our brow,
but in your love you never cease to watch over us;
Look upon the land and all who tend it at this time of crisis.
We pray for…

As we stand before you in prayer,
we pray for all who are suffering.
We acknowledge our dependence on your creation for our food,
and ask that your compassion remain with us and with all for whom we pray;
in the name of Jesus Christ our Lord.

*All* **Amen.**

## *Prayers in times of extreme weather*

1 O God of heaven and earth,
whose Son commanded the elements at the sea of Galilee,
we pray to you now as we witness weather that threatens the land,
the plants and animals we share it with and depend upon for our food.
Send us, we ask you, in this time of need,
[relief from wind and storm]
[relief from rain and flood]
[such moderate rain and showers]
that the land may recover
and we with all creation may enjoy the fruits of the earth,
to our comfort and to your praise,
through Jesus Christ our Lord.

*All* **Amen.**

2 God our creator, create in us a new heart,
a new understanding, a new way of living.
God our Redeemer, rescue us from the brink of disaster,
open our eyes to see the dangers that lie ahead.
God our Sustainer, equip and energize us,
to hold fast to the changes that we need to make.

*All* **Amen.**

## *A prayer for the waters and the seas*

Let the heavens and the earth praise you, O Lord:

*All* **the seas and all that move in them.** *Psalm 69.36*

### Readings

| | |
|---|---|
| Isaiah 43.15–19 | The Lord makes a way in the sea |
| Mark 6.45–51 | The Lord of nature |
| John 21.1–8 | The risen Lord appears on the shore of the sea of Galilee |

**Psalm 107.23–32**

23 Those who go down to the sea in ships
and ply their trade in great waters,

24 These have seen the works of the Lord
and his wonders in the deep.

25 For at his word the stormy wind arose
and lifted up the waves of the sea.

26 They were carried up to the heavens
and down again to the deep;
their soul melted away in their peril.

27 They reeled and staggered like a drunkard
and were at their wits' end.

28 Then they cried to the Lord in their trouble,
and he brought them out of their distress.

29 He made the storm be still
and the waves of the sea were calmed.

30 Then were they glad because they were at rest,
and he brought them to the haven they desired.

*All* **Let us give thanks to the Lord for his goodness,**
**and for the wonders he does for his children.**
**Let us exalt him in the congregation of the people,**
**and praise him in the council of elders.**

Eternal Father, strong to save,
whose Spirit hovered over the deep
and filled it with innumerable living things:
may our seas, lakes, and rivers teem with life;
be kept pure and free from pollution.
May all those who move through the waters
wonder at its marvels, be protected from its perils,
and be brought to the haven they desire,
through Jesus Christ our Lord.

*All* **Amen.**

## *An act of commitment for the care of creation*

'While all creatures stand in expectation, what will be the result of our liberty?'

*(Thomas Traherne)*

As the whole of creation looks with eager longing for the redemption of humankind, let us pledge ourselves anew to serve our Creator God,
the Father who is the maker of all things,
the Son through whom all things are made, and the Holy Spirit,
the giver of life, who renews the face of the earth.

Let us stand to affirm our commitment to care actively for God's creation.

*All* **Lord of life and giver of hope,**
**we pledge ourselves to care for creation,**
**to reduce our waste,**
**to live sustainably,**
**and to value the rich diversity of life.**
**May your wisdom guide us,**
**that life in all its forms may flourish,**
**and may be faithful in voicing creation's praise.**

May the commitment we have made this day be matched by our faithful living.

*All* **Amen. Amen. Amen.**

# EUCHARIST

From earliest times Christians have gathered together to celebrate the Eucharist. Indeed, the Acts of the Apostles suggests that the 'breaking of bread' was the defining practice of the first Christian community. The word eucharist literally means 'thanksgiving', and thanks are made over bread and wine, the fruit of the earth and of the vine. The thanksgiving prayer, here as elsewhere, has a trinitarian shape as we give thanks for the work of the triune God in and through creation, setting it free, and making all things new.

In the order of service provided here, we recognise that the earth is God's gift, and that we ourselves, being made in the image of God, are called to care for this fragile planet earth, to share its resources justly and to protect the variety of life it sustains. And so, in the Prayers of Penitence, provision is made for us to acknowledge our share in the plundering of natural resources and the spoiling of the natural world. Likewise, forms of Intercession are provided that focus on the theme of creation, and help us to bring before God the needs of the world at this time of ecological crisis. At the Dismissal, a form of corporate commitment is provided so that the whole congregation can express a desire to increase its awareness of the environment and strengthen its resolve to care for creation.

This Eucharist may be celebrated in the usual place of worship. But local churches are encouraged to consider arranging to have this celebration outside, either in a rural setting, or in a green space in an urban environment, according to the context of the local church.

## The Gathering

### *The Greeting*

We gather in praise of the God who is Father, Son and Holy Spirit.

*All* **Amen.**

This is the day that the Lord has made.

*All* **Let us rejoice and be glad in it.**

The Lord be with you

*All* **and also with you.**

*Silence is kept.*

## *Prayer of Preparation*

*All* **Almighty God,**
**to whom all hearts are open,**
**all desires known,**
**and from whom no secrets are hidden:**
**cleanse the thoughts of our hearts**
**by the inspiration of your holy Spirit,**
**that we may perfectly love you,**
**and worthily magnify your holy name;**
**through Christ our Lord.**
**Amen.**

## *Prayers of Penitence*

Let us confess our greed
and repent of the ways in which we have damaged God's
good creation:

We confess to you
our lack of care for the world you have given us.
Lord have mercy.

*All* **Lord have mercy.**

We confess to you
our selfishness in not sharing the earth's bounty fairly.
Christ have mercy.

*All* **Christ have mercy.**

We confess to you
our failure to protect the environment.
Lord have mercy.

*All* **Lord have mercy.**

*The president declares God's forgiveness.*

May the Father of all mercies
cleanse you from your sins,
and restore you in his image
to the praise and glory of his name,
through Jesus Christ our Lord.

*All* **Amen.**

## *Gloria in Excelsis*

*The Gloria in excelsis may be used*

*All* **Glory to God in the highest,**
**and peace to his people on earth.**
**Lord God, heavenly King,**
**almighty God and Father,**
**we worship you, we give you thanks,**
**we praise you for your glory.**
**Lord Jesus Christ, only Son of the Father,**
**Lord God, Lamb of God,**
**you take away the sin of the world:**
**have mercy on us;**
**you are seated at the right hand of the Father:**
**receive our prayer.**
**For you alone are the Holy One,**
**you alone are the Lord,**
**you alone are the Most High, Jesus Christ,**
**with the Holy Spirit,**
**in the glory of God the Father.**
**Amen.**

## *The Collect*

In a time of silence,
let us pray that our worship will bring us closer to the one who has created all things.

Glorious God,
the whole creation proclaims your marvellous work:
increase in us a capacity to wonder and delight in it,
that heaven's praise may echo in our hearts
and our lives be spent as good stewards of the earth,
through Jesus Christ our Lord.

*All* **Amen.**

## The Liturgy of the Word

### *Reading(s)*

*Either one or two readings from Scripture on pages 120–121 precede the Gospel reading.*

*At the end of each the reader may say*

This is the word of the Lord.

*All* **Thanks be to God.**

*The psalm or canticle follows the first reading; other hymns and songs may be used between the readings. Alternatively the Lament on pages 44–45 may be used.*

### *Gospel Reading*

*This acclamation may herald the Gospel reading.*

Alleluia, alleluia.
You are worthy, our Lord and God,
to receive glory and honour and power,
for you created all things.

*All* **Alleluia.**

*When the Gospel is announced the reader says*

Hear the Gospel of our Lord Jesus Christ according to *N.*

*All* **Glory to you, O Lord.**

*At the end*

This is the Gospel of the Lord.

*All* **Praise to you, O Christ.**

### *Sermon*

## *Affirmation of Faith*

*On Sundays and Principal Holy Days an authorized translation of the Nicene Creed is used, or on occasion the Apostles' Creed or an authorized Affirmation of Faith may be used.*

*All* **We believe in one God,**
**the Father, the Almighty,**
**maker of heaven and earth,**
**of all that is,**
**seen and unseen.**
**We believe in one Lord, Jesus Christ,**
**the only Son of God,**
**eternally begotten of the Father,**
**God from God, Light from Light,**
**true God from true God,**
**begotten, not made,**
**of one Being with the Father;**
**through him all things were made.**
**For us and for our salvation he came down from heaven,**
**was incarnate from the Holy Spirit and the Virgin Mary**
**and was made man.**
**For our sake he was crucified under Pontius Pilate;**
**he suffered death and was buried.**
**On the third day he rose again**
**in accordance with the Scriptures;**
**he ascended into heaven**
**and is seated at the right hand of the Father.**
**He will come again in glory to judge the living and the dead,**
**and his kingdom will have no end.**
**We believe in the Holy Spirit,**
**the Lord, the giver of life,**
**who proceeds from the Father and the Son,**
**who with the Father and the Son is worshipped and glorified,**
**who has spoken through the prophets.**
**We believe in one holy catholic and apostolic Church.**
**We acknowledge one baptism for the forgiveness of sins.**
**We look for the resurrection of the dead,**
**and the life of the world to come.**
**Amen.**

### *Intercessions*

*One of the forms on pages 65–72 may be used. This response may be used*

God of life:
*All* **hear our prayer.**

*The prayers end*

Merciful Father,
*All* **accept these prayers**
**for the sake of your Son,**
**our Saviour Jesus Christ.**
**Amen.**

## The Liturgy of the Sacrament

### *The Peace*

You shall go out in joy and be led back in peace;
the mountains and the hills before you shall burst into song,
and all the trees of the field shall clap their hands.

The peace of the Lord be always with you
*All* **and also with you.**

Let us offer one another a sign of peace.

*All may exchange a sign of peace.*

### *Preparation of the Table*

### *Taking of the Bread and Wine*

*A hymn may be sung.*

*The gifts of the people may be gathered and presented.*

*The table is prepared and bread and wine are placed upon it.*

*This prayer at the preparation of the table may be said.*

As the grain once scattered in the fields
and the grapes once dispersed on the hillside
are now united on this table in bread and wine,
so, Lord, may your whole Church soon be gathered together
from the corners of the earth
into your kingdom.

*All* **Amen.**

*The president takes the bread and wine.*

## *The Eucharistic Prayer (G)*

The Lord be with you

*All* **and also with you.**

Lift up your hearts.

*All* **We lift them to the Lord.**

Let us give thanks to the Lord our God.

*All* **It is right to give thanks and praise.**

Blessed are you, Lord God,
our light and our salvation;
to you be glory and praise for ever.

From the beginning you have created all things
and all your works echo the silent music of your praise.
In the fullness of time you made us in your image,
the crown of all creation.

You give us breath and speech, that with angels and archangels
and all the powers of heaven
we may find a voice to sing your praise:

*All* **Holy, holy, holy Lord,**
**God of power and might,**
**heaven and earth are full of your glory.**
**Hosanna in the highest.**

**Blessed is he who comes in the name of the Lord.**
**Hosanna in the highest.**

How wonderful the work of your hands, O Lord.
As a mother tenderly gathers her children,
you embraced a people as your own.
When they turned away and rebelled
your love remained steadfast.

From them you raised up Jesus our Saviour, born of Mary,
to be the living bread,
in whom all our hungers are satisfied.
He offered his life for sinners,
and with a love stronger than death
he opened wide his arms on the cross.

On the night before he died,
he came to supper with his friends
and, taking bread, he gave you thanks.
He broke it and gave it to them, saying:
Take, eat; this is my body which is given for you;
do this in remembrance of me.

At the end of supper, taking the cup of wine,
he gave you thanks, and said:
Drink this, all of you; this is my blood of the new covenant,
which is shed for you and for many for the forgiveness of sins.
Do this, as often as you drink it, in remembrance of me.

Jesus Christ is Lord:

*All* **Lord, by your cross and resurrection**
**you have set us free.**
**You are the Saviour of the world.**

Father, we plead with confidence
his sacrifice made once for all upon the cross;
we remember his dying and rising in glory,
and we rejoice that he intercedes for us at your right hand.

Pour out your Holy Spirit as we bring before you
these gifts of your creation;
may they be for us the body and blood of your dear Son.

As we eat and drink these holy things in your presence,
form us in the likeness of Christ,
and build us into a living temple to your glory.

[Remember, Lord, your Church in every land.
Reveal her unity, guard her faith,
and preserve her in peace ...]

Bring us at the last with *[N and]* all the saints
to the vision of that eternal splendour
for which you have created us;
through Jesus Christ, our Lord,
by whom, with whom, and in whom,
with all who stand before you in earth and heaven,
we worship you, Father almighty, in songs of everlasting praise:

*All* **Blessing and honour and glory and power
be yours for ever and ever.
Amen.**

### *The Lord's Prayer*

As our Saviour taught us, so we pray

*All* **Our Father in heaven,
hallowed be your name,
your kingdom come,
your will be done,
on earth as in heaven.
Give us today our daily bread.
Forgive us our sins
as we forgive those who sin against us.
Lead us not into temptation
but deliver us from evil.
For the kingdom, the power,
and the glory are yours
now and for ever.
Amen.**

### *Breaking of the Bread*

*The president breaks the consecrated bread*

We break this bread
to share in the body of Christ.

*All* **Though we are many, we are one body,
because we all share in one bread.**

*The Agnus Dei may be used as the bread is broken*

*All* **Jesus, Lamb of God, have mercy on us.**
**Jesus, bearer of our sins, have mercy on us.**
**Jesus, redeemer of the world, grant us peace.**

*or*

*All* **Lamb of God, you take away the sin of the world,**
**have mercy on us.**
**Lamb of God, you take away the sin of the world,**
**have mercy on us.**
**Lamb of God, you take away the sin of the world,**
**grant us peace.**

## *Giving of Communion*

God's holy gifts
for God's holy people.

*or*

Draw near with faith
*All* **Jesus Christ is holy,**
**Jesus Christ is Lord,**
**to the glory of God the Father.**

*The president and people receive communion. Authorized words of distribution are used and the communicant replies Amen.*

*During the distribution hymns and anthems may be sung.*

## *Prayer after Communion*

*Silence is kept.*

Creator God,
send your Holy Spirit to renew this living world,
that the whole creation,
in its groaning and striving,
may know your loving purpose
and come to reflect your glory;
in Jesus Christ our Lord.
*All* **Amen.**

## The Dismissal

*A hymn may be sung.*

*The following Act of Commitment may be used.*

'While all creatures stand in expectation, what will be the result of our liberty?'

*(Thomas Traherne)*

As the whole of creation looks with eager longing for the redemption of humankind, let us pledge ourselves anew to serve our Creator God, the Father who is the maker of all things, the Son through whom all things are made, and the Holy Spirit, the giver of life, who renews the face of the earth.

Let us stand to affirm our commitment to care actively for God's creation.

*All* **Lord of life and giver of hope,**
**we pledge ourselves to care for creation,**
**to reduce our waste,**
**to live sustainably,**
**and to value the rich diversity of life.**
**May your wisdom guide us,**
**that life in all its forms may flourish,**
**and may be faithful in voicing creation's praise.**

*The president may use this or another suitable blessing* *(see page 75)*

May God the Father of our Lord Jesus Christ,
the source of all goodness,
pour his blessing upon all things created,
and upon us his children,
that we may use his gifts to his glory
and the welfare of all peoples;
and the blessing …

*A minister says*

Go in peace, and may the commitment we have made this day be matched by our faithful living.

*All* **Thanks be to God. Amen. Amen. Amen.**

# ALTERNATIVE AND SEASONAL MATERIAL

## Invitations to Confession

1 Human sin disfigures the whole creation,
which groans with eager longing for God's redemption.
We confess our sin in penitence and faith.

2 As those who know the generosity of God,
let us confess our sins,
especially the ways in which we take creation and God's gifts for granted.

## Kyrie Confessions

1 Lord, you delight in creation, its colour and diversity;
yet we have misused the earth
and plundered its resources for our own selfish ends.
Lord, have mercy.
**Lord, have mercy.**

You have brought order out of chaos,
light in darkness, good out of evil,
but we have preferred darkness
in words and deeds which dishonour God's holy name.
Christ, have mercy.
**Christ, have mercy.**

You have showered us with blessings,
but we have been grudging towards others
and lacking in generosity in word and deed.
Lord, have mercy.
**Lord, have mercy.**

2 Lord, you give us this good earth,
yet we take your generous gifts for granted.
Lord, have mercy.
**Lord, have mercy.**

Lord, you give us this good earth,
but we squander its rich resources.
Christ, have mercy.
**Christ, have mercy.**

Lord, you give us this good earth,
but we fail to share your bounty with all of your children.
Lord, have mercy.
**Lord, have mercy.**

3 God has blessed us,
but still God's children go hungry.
Lord, have mercy.
**Lord, have mercy.**

God has blessed us,
but still the poor cry out for justice.
Christ, have mercy.
**Christ, have mercy.**

God has blessed us,
but still we see inequality and oppression in the earth.
Lord, have mercy.
**Lord, have mercy.**

4 Lord, you give us seedtime and harvest,
forgive us for those times when we waste our food.
Lord, have mercy.
**Lord, have mercy.**

Lord, you make us one bread, one body,
forgive us when we fail to share the good things you give us.
Christ, have mercy.
**Christ, have mercy.**

Lord, you give us our daily bread,
may we hunger and thirst for you, the living bread.
Lord, have mercy.
**Lord, have mercy.**

## Prayers of Penitence

1 We confess our sin, and the sins of our society,
in the misuse of God's creation.

God our Father, we are sorry
for the times when we have used your gifts carelessly,
and acted ungratefully.
Father, in your mercy:
**forgive us and help us.**

We enjoy the fruits of the harvest,
but sometimes forget that you have given them to us.
Father, in your mercy:
**forgive us and help us.**

We belong to a people who are full and satisfied,
but ignore the cry of the hungry.
Father, in your mercy:
**forgive us and help us.**

We are thoughtless,
and do not care enough for the environment and the world you have made.
Father, in your mercy:
**forgive us and help us.**

2 Creator God, maker of heaven and earth,
we acknowledge our failure to live responsibly
as part of your creation.

We have taken what we want,
without considering the consequences;
we have wasted and discarded,
without thought for the future.
Open our hearts and minds to the signs of our times,
to the groaning of creation,
so that we may turn from our greed and lack of vision
and see a world being made anew in Jesus Christ our Lord.

*All* **Amen.**

## Collects

1 Almighty God,
you have created the heavens and the earth
and made us in your own image:
teach us to discern your hand in all your works
and your likeness in all your children;
through Jesus Christ your Son our Lord,
who with you and the Holy Spirit reigns supreme over all things,
now and for ever.
**Amen.**

2 Almighty God and Father,
you have so ordered our life
that we are dependent on one another:
prosper those engaged in commerce and industry
and direct their minds and hands
that they may rightly use your gifts in the service of others;
through Jesus Christ your Son our Lord,
who is alive and reigns with you,
in the unity of the Holy Spirit,
one God, now and for ever.
**Amen.**

3 Almighty God,
whose will it is that the earth and the sea
should bear their fruit in due season:
bless the labours of those who work on land and sea,
grant us a good harvest
and the grace always to rejoice in your fatherly care;
through Jesus Christ your Son our Lord,
who is alive and reigns with you,
in the unity of the Holy Spirit,
one God, now and for ever.
**Amen.**

4 God our Father,
you never cease the work you have begun
and prosper with your blessing all human labour:
make us wise and faithful stewards of your gifts
that we may serve the common good,
maintain the fabric of our world
and seek that justice where all may share
the good things you pour upon us;
through Jesus Christ your Son our Lord,
who is alive and reigns with you,
in the unity of the Holy Spirit,
one God, now and for ever.
**Amen.**

5 Eternal God,
you crown the year with your goodness
and you give us the fruits of the earth in their season:
grant that we may use them to your glory,
for the relief of those in need and for our own well-being;
through Jesus Christ your Son our Lord,
who is alive and reigns with you,
in the unity of the Holy Spirit,
one God, now and for ever.
**Amen.**

## Gospel Acclamations

1 Alleluia, alleluia.
God has spoken to us through his Son,
through whom he created all things.
**Alleluia.**

2 Alleluia, alleluia.
The sower sows the seed, which is the word of the Lord:
Those who accept it bear fruit, thirty and sixty and a hundredfold.' *(cf Mark 4.3,8,20)*
**Alleluia.**

3 Alleluia, alleluia.
We do not live by bread alone,
but by every word that comes from the mouth of God.
**Alleluia.**

4 Alleluia, alleluia.
As the rain and the snow come down from heaven and water the earth
making it spring forth and grow,
so shall my word accomplish my purpose.
**Alleluia.**

## Intercessions

1 In peace let us pray to the Father, through the Son
and in the power of the Holy Spirit, who make, sustain and renew all things.
Heavenly Father,
we pray for your Church throughout the world
that we may be faithful to our baptism.
With the dawn of each day,
may we be awakened to the beauty of the earth
and rejoice in the wonder and diversity of creation in all its forms and colour.
Lord, hear us.
**Lord, graciously hear us.**

Lord of life,
as all living things depend on the good quality of the air, the soil, and water,
may your wisdom guide us as we care for the environment.
Deliver us from selfishness and rapacious greed.
Help us to share the rich resources of this world gladly and justly,
in the cause of stability and peace between nations and peoples.
Lord, hear us.
**Lord, graciously hear us.**

Lord of mercy,
we bring before you areas affected by chronic shortage of water,
and pray for those suffering as a result of drought
or the lack of safe water to drink.
We pray for those suffering from the effects of extreme weather
and whose environment has been damaged
by cyclones, floods or destructive wildfires.
May we better understand the effects
of the changing patterns of weather on our planet.
Lord, hear us.
**Lord, graciously hear us.**

Lord of abundant life,
we give thanks for the rich harvest of the seas;
may we cherish the good things you have created
and be successful in reducing the pollution
in our oceans, rivers and lakes that life may flourish.
May your wisdom help us to maintain the biodiversity of our fragile planet.
Strengthen our resolve and bless the efforts of all
who seek to protect the marine, animal, insect and plant life
that are threatened with extinction.
Lord, hear us.
**Lord, graciously hear us.**

Lord of creation,
you have placed us on the earth to care for it,
and call us to be co-workers with Christ your Son.
Bless our farmers, those who work in our nature reserves and National Parks,
in our Areas of Outstanding Natural Beauty and Sites of Scientific Interest.
and all who are custodians of our landscape.
Lord, hear us.
**Lord, graciously hear us.**

Lord of the universe,
you have made us from the stuff of the earth,
and to earth we shall return.
May we tread lightly upon this earth
and succeed in the innovation and development
of sources of renewable energy and green technology.
By your grace, may we live our days wisely,
live sustainably,
and at the last come with all your saints
into paradise and enjoy creation made new.

Merciful Father,
**accept these prayers**
**for the sake of your Son**
**our Saviour, Jesus Christ.**
**Amen.**

2 Let us pray to God the almighty, the Lord of creation.

God said, 'Let there be light.'
Eternal God, we thank you for your light and your truth.
We praise you for your wisdom and power
in creating this universe which proclaims your glory.
Inspire us to worship you, the creator of all,
and let your light shine upon our world.
God of life:
**hear our prayer.**

God said, 'Let there be a firmament in the midst of the heavens.'
We thank you for the vastness of the universe
and the mysteries of space.
We pray for all scientists and astronomers
who extend the boundaries of our knowledge.
As we contemplate the wonder of the heavens,
confirm us in the truth that every human being is
known and loved by you.
God of life:
**hear our prayer.**

God said, 'Let the waters be gathered together,
and let dry land appear.'
We thank you for the beauty of the earth,
for the diversity of land and sea,
for the resources of the earth.
Give us the will to safeguard this planet
and to use its riches for the good and welfare of all.
God of life:
**hear our prayer.**

God said, 'Let there be lights in the sky
to separate the day and the night.'
We thank you for the warmth of the sun,
the light of the moon, the glory of the stars.
We praise you for the formations of clouds,
the radiance of dawn and sunset.
Forgive us wasting or abusing the energy
on which all life depends.
Open our eyes to behold your beauty,
and our lips to praise your name.
God of life:
**hear our prayer.**

God said, 'Let the waters bring forth living creatures,
and let birds fly across the sky.'
We thank you for the teeming life of the seas,
and the flight of the birds.
Help us to protect the environment
so that all life may flourish.
God of life:
**hear our prayer.**

God said, 'Let us create human beings in our own image.'
We pray for the human family.
We rejoice in its diversity,
we repent of its divisions and violence.
By the power of your Spirit, restore your image in us,
through Christ who came to remake us
by his death and resurrection.
God of life:
**hear our prayer.**

Heavenly Father, you have filled the world with beauty:
open our eyes to behold your gracious hand in all your works;
that, rejoicing in your whole creation,
we may learn to serve you with gladness;
for the sake of him through whom all things were made,
your Son Jesus Christ our Lord.
**Amen.**

3 Let us pray to God,
that he will bring to fruition all that he desires
for his creation.
Father, Lord of creation,
**in your mercy, hear us.**

You have created the universe by your eternal Word,
and have blessed humankind in making us
stewards of the earth.
We pray for your world,
that we may share and conserve its resources,
and live in reverence for the creation
and in harmony with one another.
Father, Lord of creation,
**in your mercy, hear us.**

You have given the human race a rich land,
a land of streams and springs,
wheat and barley,
vines and oil and honey.
We have made by sin a world of suffering and sorrow.
We pray for those who bear the weight of affliction
that they may come to share the life of wholeness
and plenty.
Father, Lord of creation,
**in your mercy, hear us.**

In Christ you call us to a new way of life,
loving our neighbours before ourselves.
Help us to treat with care and respect the world as it is
as we live in hope and anticipation of the world
as it will be
when your kingdom comes and your will is done.
Father, Lord of creation,
**in your mercy, hear us.**

We thank you for those, living and departed,
who have shown a true respect for your creation …
Help us to follow in their footsteps,
until, with them, we see you face to face,
where all is made new in Christ our Lord.
Merciful Father,
**accept these prayers**
**for the sake of your Son**
**our Saviour, Jesus Christ.**
**Amen.**

4 Let us ask the God of Creation to send a blessing upon us.

Upon the rich earth send a blessing, O Lord.
Let the earth be fruitful
and its resources be hallowed.
We ask in faith:
**Hear us, good Lord.**

Upon human labour send a blessing, O Lord.
Prosper the work of our hands;
may all find dignity and just reward in their work;
free the exploited and oppressed.
We ask in faith:
**Hear us, good Lord.**

Upon the produce of the earth send a blessing, O Lord.
Guide us into a sustainable future,
and give us the will to share the fruits of the world.
We ask in faith:
**Hear us, good Lord.**

Upon the seas and waters send a blessing, O Lord.
Teach us to cherish the water of the earth,
and to conserve the seas, lakes and rivers.
We ask in faith:
**Hear us, good Lord.**

Upon aid agencies send a blessing, O Lord.
Where the earth is parched and the well has run dry;
where war brings want, and children go hungry;
where the poor cry out for bread and for justice,
give hands to care and heal, and compel us to be generous.
We ask in faith:
**Hear us, good Lord.**

We ask you to hear us, good Lord,
for the sake of your Son,
our Saviour Jesus Christ.
**Amen.**

5 Let us offer our prayers to God for the life of the world
and for all God's people in their daily life and work.
God, the beginning and end of all things,
in your providence and care
you watch unceasingly over all creation;
we offer our prayers
that in us and in all your people your will may be done,
according to your wise and loving purpose in Christ our Lord.
Lord of life:
**hear our prayer.**

We pray for all engaged in research to safeguard crops
against disease,
and to produce abundant life among those who hunger
and whose lives are at risk.
Prosper the work of their hands
and the searching of their minds,
that their labour may be for the welfare of all.
Lord of wisdom:
**hear our prayer.**

We pray for governments and aid agencies,
and those areas of the world affected by climate change.
By the grace of your Spirit,
touch our hearts
and the hearts of all who live in comfortable plenty,
and make us wise stewards of your gifts.
Lord of justice:
**hear our prayer.**

We offer ourselves to your service,
asking that by the Spirit at work in us
others may receive a rich harvest of love and joy and peace.
Lord of faithfulness:
**hear our prayer.**

God of grace,
as you are ever at work in your creation,
so fulfil your wise and loving purpose in us
and in all for whom we pray,
that with them and in all that you have made,
your glory may be revealed
and the whole earth give praise to you,
through Jesus Christ our Lord.
**Amen.**

## Prayers at the Preparation of the Table

1 Blessed be God,
by whose grace creation is renewed,
by whose love heaven is opened,
by whose mercy we offer our sacrifice of praise.
**Blessed be God for ever.**

2 Blessed are you, Lord God of all creation:
through your goodness we have this bread to set before you,
which earth has given and human hands have made.
It will become for us the bread of life.
**Blessed be God for ever.**

Blessed are you, Lord God of all creation:
through your goodness we have this wine to set before you,
fruit of the vine and work of human hands.
It will become for us the cup of salvation.
**Blessed be God for ever.**

3 Blessed are you, Lord God of all creation;
you bring forth bread from the earth.
**Blessed be God for ever.**

Blessed are you, Lord God of all creation;
you create the fruit of the vine.
**Blessed be God for ever.**

## Eucharistic Prefaces

1 And now we give you thanks
because in him, our risen Lord,
the new creation is being brought to perfection,
a broken world is being renewed,
and creation itself will share
in the glorious liberty of the children of God.

2 And now we give you thanks
for your ancient promise
that while the earth endures,
seedtime and harvest, cold and heat,
summer and winter, day and night,
will never cease.

3 And now we give you thanks
because all things are of your making,
all times and seasons obey your laws,
but you have chosen to create us in your own image,
setting us over the whole world in all its wonder.
You have made us stewards of your creation,
to praise you day by day
for the marvels of your wisdom and power:
so earth unites with heaven
to sing the new song of creation:
**Holy, holy, holy Lord …**

*The following prefaces are for use with Additional Eucharistic Prayer (2) 'on occasions when a significant number of children are present'.*[7]

4 You give us sun and moon and star-lit sky,
everything that gives us light,
light for our eyes, our hearts, our minds.

5 You give us the fish in the sea,
the birds of the air,
and every plant and tree;
the life that sleeps in the winter earth,
and awakens again in the spring.

## Post Communions

1 God our creator,
by your gift
the tree of life was set at the heart of the earthly paradise,
and the bread of life at the heart of your Church:
may we who have been nourished at your table on earth
be transformed by the glory of the Saviour's cross
and enjoy the delights of eternity;
through Jesus Christ our Lord.
**Amen.**

2 God our creator,
you give us seed to sow and bread to eat:
as you have blessed the fruit of our labour in this eucharist,
so we ask you to give all your children their daily bread,
that the world may praise you for your goodness;
through Jesus Christ our Lord.
**Amen.**

3 Lord of the harvest,
with joy we have offered thanksgiving for your love in creation
and have shared in the bread and the wine of the kingdom:
by your grace plant within us a reverence for all that you give us
and make us generous and wise stewards
of the good things we enjoy;
through Jesus Christ our Lord.
**Amen.**

[7] *Common Worship: Additional Eucharistic Prayers with Guidance on Celebrating the Eucharist with Children*, CHP, © Archbishops' Council 2012 p.1

## Blessings

1 May God the Father,
who clothes the lilies of the field
and feeds the birds of the air,
provide you with all we need for life in its fullness.
**Amen.**

May God the Son,
who fed the five thousand and turned water into wine,
feed you with his life and transform us in his love.
**Amen.**

May God the Holy Spirit,
who hovered over the waters of creation
and formed the world from chaos,
form you in the likeness of Christ and renew the face of the earth.
**Amen.**

And the blessing …

2 May God
who clothes the lilies of the field and feeds the birds of the air,
who leads the lambs to pasture and the deer to water,
who multiplied loaves and fishes, and changed water into wine,
lead us and feed us,
and change us to reflect the glory of our Creator
now and through all eternity;
and the blessing …

## Dismissals and Endings

1 Tend the earth, care for God's good creation,
and bring forth the fruits of righteousness.
Go in the peace of Christ.
**Thanks be to God.**

2 Freely you have received, freely give.
Go in peace to love and serve the Lord.
**In the name of Christ. Amen.**

# A SEASON OF CREATION:
## Thematic resources for use over four weeks

These three strands of thematic resources can be used during the Season of Creation in September and early October, or at other points during the year when a church community wishes to focus on creation and environmental themes. If used in September this may be enriched by an observance of Holy Cross Day (page 112) and come to focus either at Harvest Festival or the Feast of St Francis of Assisi (page 115).

## Strand 1 – Of Every Kind

This is based on the four annual observances resourced by 'Seasons and Festivals of the Agricultural Year' in *Common Worship: Times and Seasons*. These four times of the year reflect the life-giving process of sowing seed, caring for the growing crops, harvesting, eating and storing them. During the Season of Creation this natural process might be the focus of four weeks of reflection.

Blessed are you, Lord God of all creation:
*All* **For in your abundant care you have given us fertile land.**

*Collect*

Ever living God,
you created the earth
which gives us its food in due season;
give us the will to cherish the earth
and to use its riches for the good and welfare of all people.
This we ask through Jesus Christ our Lord.

*All* **Amen.**

| ***Week 1: Sowing*** | ***Week 2: Growing*** | ***Week 3: Gathering*** | ***Week 4: Treasuring*** |
|---|---|---|---|
| Based on Plough Sunday | Based on Rogationtide | Based on Lammastide | Based on Harvest |
| Genesis 1.9–13 | Ezekiel 47.[1–6a] 6b–12 | Leviticus 23.9–14 | Deuteronomy 26.1–11 |
| Psalm 104.1,6,11–26 | Psalm 147.1–13; Benedicite | Psalm 126 | Psalm 65 |
| 2 Corinthians 9.9–11a | Philippians 4.4–7 | James 5.7–9 | Colossians 1.1–15 |
| Matthew 6.25–34 *or* Matthew 13.1–9 | Matthew 6.7–15 *or* John 12.23–26 | Matthew 15.32–39 *or* John 4.31–38 | Luke 12.16–30 |

## Strand 2 – The Garden

This is based on the agricultural year themes in *Common Worship: Times and Seasons* but in a form that may be more readily accessible to people living in less rural situations, by children (e.g. in schools or congregations where there is a large proportion of children) and perhaps by those less familiar with church.

Blessed are you, Lord God of all creation:

*All* **For you give us the fruits of the earth in their season.**

*Collect*

Lord of all life,
you created the universe,
and gave us this beautiful earth
to discover and to cherish;
help us to wonder at creation,
to treasure your gifts
and to use them to your glory.

*All* **Amen.** *(cf Additional Eucharistic Prayers: Prayer 2)*

| *Week 1: The Spade* | *Week 2: The Watering Can* | *Week 3: The Basket* | *Week 4: The Plate* |
|---|---|---|---|
| Genesis 1.9–13, 29–31 | Joel 2.21–27 *or* Song of Songs 2.10–13 *or* Isaiah 55.10–13 *or* Ezekiel 47.[1–6a] 6b–12 | Deuteronomy 26.1–11 | Isaiah 55.1–2 |
| Psalm 126 | Psalm 65.8–13 | Psalm 104.1,29–32 | Psalm 107 |
| | | | Acts 2.43–47 |
| Matthew 13.1–9 *or* Matthew 13.31–33 | Luke 13.6–9 | Mark 4.26–32 | Matthew 15.32–39 *or* John 21.1–14 *or* Luke 14.15–23(24) |

## Strand 3 – Creation Emergency

This offers a way to address the climate and environmental emergency directly within the context of an understanding of Creation as a gift. Over the four weeks of the Season of Creation the readings reflect on the wonder and abundance of creation, the damage that is wreaked upon it and the desire and hope for healing.

Blessed are you, Lord God of all creation,
with you is the well of life:

*All* **And in your light shall we see light.**

## *Collect*

God our Creator,
who laid the foundations of the earth
and stretched out the heavens as a tent cloth;
how glorious is your name in all the world.
Give us grace to wonder at your gift,
to treasure the earth
and to work for its protection and renewal;
through Jesus Christ your Son, our Lord,
who is alive and reigns with you,
in the unity of the Holy Spirit,
one God, now and for ever.

***All*** **Amen.**

| ***Week 1:*** **Wonder** | ***Week 2:*** **Abundance** | ***Week 3:*** **Desolation** | ***Week 4:*** **Restoration** |
|---|---|---|---|
| Job 38.1–11,16-18 *or* Proverbs 8.1–4,22–31 | Deuteronomy 8.7–18 | Isaiah 5.1–10 *or* Jeremiah 14.1–21 | Genesis 8.12–20 *or* Genesis 9.8–17 *or* Job 42.10-17 |
| Psalm 33 *or* Psalm 89 Ecclesiasticus 42.15–33 | Psalm 19; Benedicite | Psalm 102 | Psalm 126 |
| Colossians 1.1–15 | Revelation 22.1–5 | Revelation 6.12–17 | Romans 8.18–22 |
| John 1.1–14 | Matthew 6.25–34 | Mark 11.12–20 *or* Matthew 24.15–28 | Mark 4.30–32 *or* John 6.1–13 |

# SEASONS AND FESTIVALS OF THE AGRICULTURAL YEAR

## Introduction

The Jewish and Christian Scriptures give eloquent expression to the creative power and wisdom of God. It is therefore a natural instinct for worshipping communities to develop patterns of worship and prayer around the agricultural year. Of course, there were dangers, and the same Scriptures bear witness to concerns about the idolatry of fertility cults and the worship of created things rather than the creator. Nevertheless, ancient society lived close to the land, and it is no surprise that the ancient Jewish festivals of Passover and Unleavened Bread, Weeks and Tabernacles all have agrarian roots. The Christian tradition, too, has assimilated, but with differing emphases and in different times and places, particular agricultural festivals. Much of this is bound up with the need to provide food to sustain human life, and the accompanying sense of a proper humility before God as source of all things, gratitude for his goodness, and responsibility in stewarding the resources of the earth. In more recent years, urban congregations have explored ways of adapting traditional creation-based festivals for their own contexts.[8] This provision is not intended to limit such adaptation.

### *Plough Sunday*

The observance of Plough Sunday on the First Sunday of Epiphany goes back to Victorian times, but behind it there is a much older observance, associated with the first working day after the twelve days of Christmas, hence 'Plough Monday' in some places. In days when work was scarce in winter, the observance looked forward to the time of sowing with the promise of a harvest to come. Some Christian communities have reintroduced it as a focus for asking a blessing on human labour near the start of the calendar year.

Forms are provided here for the blessing of a plough and the blessing of seed. Lectionary material is also provided.

---

[8] See ideas in, for example, *Creative Ideas for Wild Church: Taking all-age worship and learning outdoors* and *Together for a Season: All-age resources for Feasts and Festivals of the Christian Year* (details in Appendix)

## *Rogationtide*

The Rogation Days (from the Latin *rogare*, 'to ask') are the three weekdays before Ascension Day. However, in practice, many churches have observed Rogation on the preceding Sunday (Easter V in the Prayer Book, the Sixth Sunday of Easter in *Common Worship*). The Prayer Book Gospel includes the words of Jesus, 'Whatsoever ye shall ask for in my Name, he will give it you' – words associated with the heavenly intercession of the ascended Christ.

Originally, the Christian observance of Rogation was taken over from Graeco-Roman religion, where an annual procession invoked divine favour to protect crops against mildew. The tradition grew of using processional litanies, often around the parish boundaries, for the blessing of the land. These processions concluded with a mass. The Rogation procession was suppressed at the Reformation, but it was restored in 1559. The poet George Herbert interpreted the procession as a means of asking for God's blessing on the land, of preserving boundaries, of encouraging fellowship between neighbours with the reconciling of differences, and of charitable giving to the poor. The tradition of 'beating the bounds' has been preserved in some communities, while others maintain the traditional use of the Litany within worship. In more recent times, the scope of Rogation has been widened to include petition for the world of work and for accountable stewardship, and prayer for local communities, whether rural or urban.

The provision here includes resource material for Rogationtide worship, litanies for the Rogationtide procession, and patterns of readings. The material is to be used with flexibility according to local circumstances.

## *Lammastide*

Lammas or 'Loaf-mass' (derived from the Anglo-Saxon *Hlafmaesse*) is an English feast in origin, held on 1 August as a thanksgiving for the first-fruits of the wheat harvest. Traditionally, a newly baked loaf from the wheat harvest was presented before God within the mass of that day. While the ceremony ceased at the Reformation, reference to Lammas Day continued in the Prayer Book calendar, and the practice has been revived in some places in more recent years. The tradition of giving thanks for the first-fruits need not be limited to 1 August, and churches are at liberty to decide when to hold such a celebration. Here, material is provided to form the Gathering rite of the eucharist. The Lammas loaf should ideally be baked by members of the congregation, using local produce wherever possible. Other small loaves or buns, in the tradition of 'blessed bread', may be distributed to the congregation. Part of the Lammas loaf may be used as the eucharistic bread on this occasion. Two patterns of readings are suggested, the first concerning the offering of the first-fruits and the second concerning the bread of life.

### *Harvest*

Harvest Thanksgiving is a more modern addition to the church calendar. Its origins are usually traced to the adaptation in 1843 of Lammas Day by the Revd R. S. Hawker, a parish priest in Cornwall. He chose the first Sunday in October as a Christian response to coincide with the traditional but largely secular 'harvest home' celebration, but there is some evidence to suggest that a thanksgiving for the harvest was already a relatively widespread practice. An annual church celebration of the harvest certainly established itself rapidly with great popularity and was first recognized officially in the Church of England in 1862. Since then, many local traditions for the celebration have developed and many liturgical resources are available. Here, a bank of resource material is provided for use at Holy Communion or a Service of the Word. An Act of Thanksgiving is provided, which may accompany the tradition of bringing to church gifts of fresh produce and other foodstuffs.

# Plough Sunday

## The Blessing of the Plough & the Blessing of Seed

### *Note*

Forms of prayer are provided for the blessing of a plough and the blessing of seed, along with lectionary provision. The material can be incorporated into Holy Communion or a Service of the Word. Other resource material from Seasons and Festivals of the Agricultural Year may be used or adapted appropriately.

### *The Blessing of the Plough or Drill*

Blessed are you, Lord God of all creation:
for in your abundant care you have given us fertile land,
rich soil, the seasons in their courses.
You provide seed for sowing, water, light and warmth
to bring forth the miracle of growth.
You give us skill to work the land,
to prepare and nourish it, that it may be fruitful.
By your blessing,
let this plough be a sign of all that you promise to us.
Prosper the work of our hands,
and provide abundant crops for your people to share.
Blessed be God, Father, Son and Holy Spirit.
**Blessed be God for ever.**

God speed the plough.
**God speed the plough.**

### *The Blessing of Seed*

Blessed are you, Lord God of all creation:
in your goodness you have given us this seed to sow.
In it we perceive the promise of life,
the wonders of your creative love.
By your blessing,
let this seed be for us a sign of your creative power,
that in sowing and watering,
tending and watching,
we may see the miracle of growth,
and in due course reap a rich harvest.

As this seed must die to give life,
reveal to us the saving power of your Son,
who died that we might live,
and plant in us the good seed of your word.
Blessed be God, Father, Son and Holy Spirit.
**Blessed be God for ever.**

By itself the earth produces:
**first the stalk, then the ear, then the full grain shall appear.**

*The Collect*

Almighty God and Father,
you have so ordered our life
that we are dependent on one another:
prosper those engaged in commerce and industry
and direct their minds and hands
that they may rightly use your gifts in the service of others;
through Jesus Christ your Son our Lord,
who is alive and reigns with you,
in the unity of the Holy Spirit,
one God, now and for ever.
***All*** **Amen.**

*Readings*

Genesis 1.9–13 *or* Genesis 8.20—9.3 *or* Isaiah 55.6–11 *or*
Ecclesiasticus 38.25–end
Psalm 37.22–38 *or* Psalm 104
1 Corinthians 9.6–14
Matthew 6.25–end

# Rogation

## The Rogationtide Procession

### *Note*

This may be a procession within the church building or in the open air, whether the traditional beating of the bounds or a specified procession in part of the parish. Prayers and readings may be used at suitable places; in urban parishes, these might include street corners, parks and open spaces, and commercial or industrial sites.

The following resources may be used according to circumstances.

### A

*All or parts of the following litany may be said or sung*

**I**

Let us pray.

God the Father, Lord of creation,
**have mercy upon us.**

God the Son, through whom all things were made,
**have mercy upon us.**

God the Holy Spirit, who renews the face of the earth,
**have mercy upon us.**

Holy, blessed and glorious Trinity, creating and saving God,
**have mercy upon us.**

Remember, Lord, your mercy and loving-kindness towards us.
Bless this good earth, and make it fruitful.
Bless our labour, and give us all things needful for our daily lives.
Bless the homes of this parish and all who live within them.
Bless our common life and our care for our neighbour.
**Hear us, good Lord.**

**II**

For all cities, towns and villages,
and for their well-being and prosperity,
let us pray to the Lord.
**Lord, have mercy.**

For the rural economy and for its regeneration,
let us pray to the Lord.
**Lord, have mercy.**

For those who tend the countryside
and preserve its order and beauty,
let us pray to the Lord.
**Lord, have mercy.**

For traditional rural skills and crafts
and for those who exercise them,
let us pray to the Lord.
**Lord, have mercy.**

For all farms, all who work them,
and for the whole farming industry,
let us pray to the Lord.
**Lord, have mercy.**

For those who make farming policy,
and for all with authority in government,
let us pray to the Lord.
**Lord, have mercy.**

**III**

For a blessing on our land we pray.
**Hear us, good Lord.**

For healthy crops and abundant harvests we pray.
**Hear us, good Lord.**

For the care and welfare of animals
and for the veterinary profession we pray.
**Hear us, good Lord.**

For the harvest of the soil
and for the fruits of the earth in their seasons we pray.
**Hear us, good Lord.**

For seasonable weather we pray.
**Hear us, good Lord.**

For protection from blight, pestilence and disease we pray.
**Hear us, good Lord.**

For those engaged in agricultural research we pray.
**Hear us, good Lord.**

For the service industries that support rural life we pray.
**Hear us, good Lord.**

For the ministry of your Church in rural areas we pray.
**Hear us, good Lord.**

For parts of the world where the harvests have failed we pray.
**Hear us, good Lord.**

For charities, aid agencies and overseas development we pray.
**Hear us, good Lord.**

## IV

For our daily bread:
**we pray to you, O Lord.**

For all who work on the land to bring us our food in due season:
**we pray to you, O Lord.**

For all who fish the rivers, lakes and seas:
**we pray to you, O Lord.**

For all who process foods and prepare them for distribution and sale:
**we pray to you, O Lord.**

For all supermarkets and shops, and for all who work in them:
**we pray to you, O Lord.**

For those who work in food research:
**we pray to you, O Lord.**

For those who distribute food to those in need:
**we pray to you, O Lord.**

For the will to share your bounteous gifts:
**we pray to you, O Lord.**

## V

For the world of work in all its diversity:
**hear us, good Lord.**

For the industry and workplaces of this *parish/community*:
**hear us, good Lord.**

For the right ordering of work in time of technological change:
**hear us, good Lord.**

For communities that have lost traditional industries,
and for their regeneration:
**hear us, good Lord.**

For all expanding industries and for the promise of new jobs:
**hear us, good Lord.**

For small businesses and co-operatives:
**hear us, good Lord.**

For local trade and commerce:
**hear us, good Lord.**

For all service industries that provide for our daily needs:
**hear us, good Lord.**

For the unemployed and for those living in poverty:
**hear us, good Lord.**

For school leavers and all who are seeking to enter employment:
**hear us, good Lord.**

For the retired and those unable to work:
**hear us, good Lord.**

For all who work as volunteers:
**hear us, good Lord.**

**VI**

Almighty and everlasting God,
you are always more ready to hear than we to pray
and to give more than either we desire or deserve:
pour down upon us the abundance of your mercy,
forgiving us those things of which our conscience is afraid
and giving us those good things
    which we are not worthy to ask
but through the merits and mediation
of Jesus Christ your Son our Lord.
**Amen.**

**B**

*Collect*

Almighty God,
whose will it is that the earth and the sea
should bear their fruit in due season:
bless the labours of those who work on land and sea,
grant us a good harvest
and the grace always to rejoice in your fatherly care;
through Jesus Christ your Son our Lord,
who is alive and reigns with you,
in the unity of the Holy Spirit,
one God, now and for ever.
**Amen.**

*The following readings may be used at suitable points in the procession. A selection may be made from either Table A or Table B, according to circumstances. Suitable hymns or songs may be used in place of the psalmody.*

| | A | | B |
|---|---|---|---|
| 1 | Deuteronomy 26.1–11 | Psalm 147.1–12 | Genesis 8.13–end |
| 2 | Job 38.1–11,16–18 | Psalm 104.25–end | Leviticus 26.1–13 |
| 3 | Ezekiel 47.[1–6] 7–12 | Psalm 107.23–32 | Deuteronomy 8.1–10 |
| 4 | Ecclesiasticus 38.27–32 | Psalm 107.1–9 | Hosea 2.18–end |
| 5 | Acts 14.8–18 | Psalm 107.33–end | James 4.7–11 |
| 6 | 2 Corinthians 9.9–11a | Psalm 112 | Romans 8.18–25 |
| 7 | Philippians 4.4–7 | Benedicite | 1 Corinthians 3.10–14 |
| 8 | Mark 11.22–24<br>or Matthew 6.7–15 | | John 6.22–40<br>*or* John 12.23–26 |

# Lammas

## The Presentation of the Lammas Loaf

### *Note*

As the word 'Lammas' suggests, the setting for this material is the Holy Communion. the Lammas loaf, or part of it, may be used as the bread of the Eucharist, or the Lammas loaf and the eucharistic bread may be kept separate. To avoid any possible confusion, the rite of the presentation of the Lammas loaf should form part of the Gathering in Holy Communion Order One. Suitable patterns of readings are suggested.

## The Gathering

*At the entry of the ministers a hymn may be sung.*

*The president may say*

In the name of the Father,
and of the Son,
and of the Holy Spirit.

*All* **Amen.**

**The Greeting**

*The president greets the people*

The Lord be with you

*All* **And also with you.**

**The Presentation of the Lammas Loaf**

*The president introduces the service with these or other suitable words*

Brothers and sisters in Christ, the people of God in ancient times presented to the Lord an offering of first-fruits as a sign of their dependence upon God for their daily bread. At this Lammastide, we bring a newly baked loaf as our offering in thanksgiving to God for his faithfulness.

Jesus said, 'I am the bread of life; those who come to me shall never be hungry and those who believe in me shall never thirst.' *John 6.35*

*The Lammas loaf is brought to the president.*

Blessed are you, Lord God of all creation;
you bring forth bread from the fields
and give us the fruits of the earth in their seasons.
Accept this loaf, which we bring before you,
made from the harvest of your goodness.
Let it be for us a sign of your fatherly care.
Blessed are you, Lord our God,
worthy of our thanksgiving and praise.

***All*** **Blessed be God for ever.**

*The service continues with the Prayers of Penitence.*

*The following Kyrie may be used*

Lord, you give us seedtime and harvest,
forgive us for those times when we waste our food.
Lord, have mercy.

***All*** **Lord, have mercy.**

Lord, you make us one bread, one body,
We are sorry that we fail to share the good things you give us.
Christ, have mercy.

***All*** **Christ have mercy.**

Lord, you provide our daily bread,
may we hunger and thirst for you, the living bread.
Lord, have mercy,

***All*** **Lord, have mercy.**

### Readings

*The Offering of First-fruits*
Leviticus 23.9–14
Psalm 126
1 Corinthians 15.12–20 *or* James 5.7–9
Matthew 15.32–39 *or* John 4.31–38

*The Bread of Life*
Exodus 16.11–15
Psalm 104.10–15
2 Corinthians 9.6–11
John 6.28–35

### Collect

God our Father,
you never cease the work you have begun
and prosper with your blessing all human labour:
make us wise and faithful stewards of your gifts
that we may serve the common good,
maintain the fabric of our world
and seek that justice where all may share
the good things you pour upon us;
through Jesus Christ your Son our Lord,
who is alive and reigns with you,
in the unity of the Holy Spirit,
one god, now and for ever.
**Amen.**

# Harvest

## Thanksgiving for the Harvest & The Bringing Forward of Symbols of the Harvest

### Thanksgiving for the Harvest

Let us give thanks to God,
the God of all peoples of the earth.

For the colour and forms of your creation
and our place within it,
we bring our thanks, good Lord:
**your mercy endures for ever.**

For our daily food,
and for those whose work and skill
bring your good gifts to us,
we bring our thanks, good Lord:
**your mercy endures for ever.**

For the gifts and graces inspired in human minds and hearts;
for insight and imagination,
for the skills of research
which bring healing and fulfilment to the lives of many;
we bring our thanks, good Lord:
**your mercy endures for ever.**

For the light and shades of the changing seasons,
and their variety and dependability;
for new life and growth out of barrenness and decay;
we bring our thanks, good Lord:
**your mercy endures for ever.**

For new hope and strength in our communities,
especially in your Church and among all you call to serve you,
we bring our thanks, good Lord:
**your mercy endures for ever.**

For all in whose lives we see
goodness, kindness, gentleness, patience and humility,
and all the fruit of the Spirit,
we bring our thanks, good Lord:
**your mercy endures for ever.**

For the life we have been given,
and for all those whom you have given us to share it,
we bring our thanks, good Lord:
**your mercy endures for ever.**

### The Bringing Forward of Symbols of the Harvest

Let us bring forward symbols of the harvest,
gifts that God has created and his sun and rain have nurtured.
**Thanks be to God.**

Bring forward the harvest of the cornfields,
the oats and the wheat, the rye and the barley.
**Thanks be to God.**

Bring forward the harvest of roots,
the swedes and mangolds, turnips and sugar beet.
**Thanks be to God.**

Bring forward the harvest of seeds for next year's crops,
for clover, for hay and for corn.
**Thanks be to God.**

Bring forward the harvest of vegetables,
peas, potatoes, beans and hops.
**Thanks be to God.**

Bring forward the harvest of pears and apples, berries and herbs.
**Thanks be to God.**

Bring forward the harvest of flowers,
the finest blooms from our gardens and our fields.
**Thanks be to God.**

Bring forward the grain and the grape,
for our Saviour took bread and wine
to feed us with his body and his blood,
given and shed for the life of the world.
Let us feed on him by faith with thanksgiving.
**Thanks be to God.**

**Praise God, from whom all blessings flow,**
**praise him, all creatures here below,**
**praise him above, ye heavenly host,**
**praise Father, Son, and Holy Ghost.**
**Amen.**

## Collect

Eternal God,
you crown the year with your goodness
and you give us the fruits of the earth in their season:
grant that we may use them to your glory,
for the relief of those in need and for our own well-being;
through Jesus Christ your Son our Lord,
who is alive and reigns with you,
in the unity of the Holy Spirit,
one God, now and for ever.
**Amen.**

# FESTIVALS AND LESSER FESTIVALS

## The Festival of Creation in Christ (First Sunday in September)

**Festival – White or Green**

### *Invitation to Confession*

The earth is the Lord's and all that fills it. Let us confess our failure to receive with gratitude God's gift of creation, and repent of the ways in which we have caused it harm.

### *Kyrie Confession*

You laid the foundations of the earth,
but we have claimed your world as our own.
Lord, have mercy.

*All* **Lord, have mercy.**

All things were created through you and for you,
but we have not lived as those who belong to you.
Christ, have mercy.

*All* **Christ, have mercy.**

You make all things new,
but our hope has grown dim.
Lord, have mercy.

*All* **Lord, have mercy.**

### *Collect*

Holy and eternal God,
you are the Maker of heaven and earth,
you are the Word through whom all things were made,
and you are the Spirit, the giver of life:
draw us into the mystery of your love,
that with all creation we may proclaim your praise,
through Jesus Christ our Lord.

*or*

O God and Father of all that was, is and ever shall be,
help us to be thankful for all that is made
at the hands of your Son
and through the wisdom of your Spirit,
that we may care in heart and in deed for the life of our world,
to the praise and glory of your holy name.

*Lectionary*

**Evening Prayer on the Eve (if required)**
Psalm 19
Genesis 1.1–5, 26–31
Hebrews 1.1–4

**Principal Service**
Isaiah 55.12–end
Psalm 96
Colossians 1.15–20
John 1.1–14
*or* Matthew 8.23–27

**Second Service**
Ezekiel 47.1–12
*or* Isaiah 45.2–8
**Evening Psalm** 104
Revelation 22.1–5
**HC** John 15.1–17

**Third Service**
Job 38.1–11,16–18
*or* Sirach 18.1–14
*or* Sirach 43.9–12, 23–26, 28–33
**Morning Psalm** 148
Revelation 4
*or* 2 Peter 3.8–13

*Gospel Acclamation*

Alleluia, alleluia.
You are worthy, our Lord and God,
to receive glory and honour and power,
for you created all things.

*All* **Alleluia.**

*Introduction to the Peace*

You shall go out in joy and be led back into peace;
the mountains and the hills before you shall burst into song,
and all the trees of the field shall clap their hands.

*Isaiah 55.12*

*Prayer at the Preparation of the Table*

Blessed are you, Lord God of all creation;
by your goodness the world was formed and filled with plenty.
As we set before you this bread and wine,
teach us gratitude for your gifts
and make us faithful stewards of your creation.

*All* **Blessed be God for ever.**

or

As the grain once scattered in the fields
and the grapes once dispersed on the hillside
are now united on this table in bread and wine,
so, Lord, may your whole Church soon be gathered together
from the corners of the earth
into your kingdom.

*All* **Amen.**

*Preface*

And now we give you thanks
because in him, our risen Lord,
the new creation is being brought to perfection,
a broken world is being renewed,
and creation itself will share
in the glorious liberty of the children of God.

### *Extended Preface*

It is indeed right and good,
always and everywhere to give you thanks,
holy Father, almighty and eternal God,
through Jesus Christ your Son our Lord.

For by your word the heavens were made,
and all their host by the breath of your mouth.
You formed us in your image, the crown of your creation,
and entrusted us with the works of your hands.

In your great goodness you sent your Son,
the firstborn of all creation,
who holds all things together in himself,
and through whom you make all things new.

You have given us the first-fruits of the Spirit,
as creation itself waits with eager longing
to obtain the glorious freedom of the children of God.

And so, with angels and archangels
and with all that you have made,
we join the song of your glory,
forever praising you and saying:

### *Post Communion*

God our Creator,
by your gift the tree of life was set
    at the heart of the earthly paradise,
and the bread of life at the heart of your Church:
may we who have been nourished at your table on earth
be transformed by the glory of the Saviour's cross
and enjoy the delights of eternity;
through Jesus Christ our Lord.

***All*** **Amen.**

*Before the Dismissal the Act of Commitment for the Care of Creation (see page 48) may be used.*

## *An act of commitment for the care of creation*

'While all creatures stand in expectation, what will be the result of our liberty?'

*(Thomas Traherne)*

As the whole of creation looks with eager longing for the redemption of humankind, let us pledge ourselves anew to serve our Creator God,
the Father who is the maker of all things,
the Son through whom all things are made, and the Holy Spirit,
the giver of life, who renews the face of the earth.

Let us stand to affirm our commitment to care actively for God's creation.

*All* **Lord of life and giver of hope,**
**we pledge ourselves to care for creation,**
**to reduce our waste,**
**to live sustainably,**
**and to value the rich diversity of life.**
**May your wisdom guide us,**
**that life in all its forms may flourish,**
**and may be faithful in voicing creation's praise.**

May the commitment we have made this day be matched by our faithful living.

*All* **Amen. Amen. Amen.**

### *Blessing*

May God the Father,
whose glory is proclaimed in the heavens,
awaken in you a delight and wonder in this good creation.

*All* **Amen.**

May God the Son,
the Word made flesh,
through whom all things were made,
fill you with his risen life.

*All* **Amen.**

May God the Holy Spirit,
who hovered over the waters of creation
and formed the world from chaos,
form you in the likeness of Christ
and renew the face of the earth.

***All*** **Amen.**

and the blessing...

### *Daily Prayer*

*At Morning Prayer, canticle 32 (A Song of the New Creation) may be said.*

*At Evening Prayer, canticle 84 (A Song of Francis of Assisi) may be said.*

| | |
|---|---|
| *Eve: Magnificat* | By your word, O God, the heavens were made,<br>and all their host by the breath of your mouth. |
| *Benedictus* | The spirit of God has made me,<br>and the breath of the Almighty gives me life. |
| *Magnificat* | You are worthy, our Lord and God,<br>to receive glory and honour and power,<br>for you created all things. |

## The Festival of Creation in Christ: Commentary by the Liturgical Commission

The Festival of Creation in Christ is a rich affirmation of our Christian faith in that it asserts that we are not self-originating or indeed self-sufficient. Our very existence is dependent upon the free, loving and continuing act of God. We are also encouraged to see the world in the whole arc of God's purposes: created through Christ, reconciled in Christ, and destined in Christ for the liberty and glory of the new creation. To confess God as Creator is therefore to say something about God, something about the world, and something about ourselves. God is the one from whom all things come, by whom all things are sustained, and towards whom all things are ordered. The world is the work of divine wisdom, generosity and delight. Human beings are made in the image and likeness of God and entrusted with a distinctive vocation within a world which belongs first and finally to its Maker, and which God has pledged to renew.

This new Festival should not be misconstrued as a general celebration of nature or as a competitor to traditional forms of thanksgiving for the harvest. It does indeed include a call to gratitude for the beauty of the natural world; a call to action to arrest ecological crisis; and an invitation to take renewed responsibility for natural resources. But the Festival *of Creation in Christ* is rightly concerned with creation as a mystery of faith: the creation of all things by God, the holding together of all things in Christ, and the promised renewal of all things in the risen Lord. This is precisely the theological emphasis of the international, ecumenical conversations which have fostered the adoption of the Festival. Theologians, liturgiologists, environmentalists, and ecologists from many Christian traditions, meeting in Assisi on several occasions in recent years, have articulated that such a feast should commemorate God's act of creation; contemplate creation as a Christological mystery; confess creation as a Trinitarian mystery; and receive creation as a sacramental and covenanted gift.[9] To these themes the Church of England's liturgical resources offer a distinctly paschal and eschatological accent: the creation celebrated by this Festival is the creation now being renewed in the risen Christ.

---

[9] *Theological Overview of the Feast of Creation: celebrating a foundational tenet of the Christian faith*, produced by the World Council of Churches' Steering Committee, with support from its theological sub-committee (bit.ly/feast-of-creation-theology)

Many worshippers will be familiar with the language of environmental care. Fewer will have been taught to see that care as flowing from the doctrine of creation itself, or to see the doctrine of creation as ordered towards new creation. The new teaching document from the House of Bishops, *Hope for All Creation* (2026), helps to make the connection. It begins with the affirmation made at the Lambeth Confence 2022 that we have been granted a world of 'breath-taking beauty, astounding abundance and intricate interconnection', a world which God declared good and loves. It then sets that affirmation within the reality of crisis: damage to climate and ecosystems is widespread and intensifying, biodiversity loss is escalating, and vulnerable communities are suffering acutely. Yet the report's theological centre is hope. God has not forsaken the Earth. God sustains the 'interconnected webs of life'; God has come in Jesus Christ 'to dwell in our creaturely home'; God has reconciled us to himself and to fellow creatures; God has raised Christ from the dead as pledge of the renewal and liberation of all creation; and God sends the Holy Spirit to enable the Church to live out that reconciliation and hope.

The Festival of Creation in Christ stands at the intersection of worship and mission. It gives the Church a liturgical means of saying what its teaching documents have long been trying to recover: that care for creation is not an optional ethical addition to Christian discipleship, but an implication of the Church's confession of God and of its hope in the resurrection. *Hope for All Creation* locates the present moment within a longer Anglican tradition of theological reflection on the environment as early as the 1941 Malvern call for renewed reverence for the earth and its resources as a storehouse of divine bounty. In that tradition, *Sharing God's Planet* (2012) framed the connectedness of all things in covenantal terms, with humanity called to prophetic witness, priestly offering and servant leadership under God's care.

The new Festival is therefore a liturgical crystallisation of a much longer theological, pastoral and ecumenical development. Anglican worship has already been shaping Christians for this vocation through the pattern of the Christian year. Advent teaches longing for renewal on a cosmic scale with the coming of Christ in glory at the end of the age; Epiphany celebrates the manifestation of Christ to the world; Lent reminds us of the holiness of restraint, fasting and simplicity; Easter, above all, proclaims resurrection and the renewal of all creation. In recent decades many Anglicans have also observed a so-called 'Season of Creation' in early autumn.

In the Byzantine tradition 1 September is associated with the beginning of the liturgical year and with the creation of the world (and the Ecumenical Patriarchate in 1989 invited all Christians to pray on that day). In recent decades, 1 September has been known across many churches as Creation Day, the Feast of Creation, or the World Day of Prayer for Creation. The Orthodox Church's historical observance of 1 September inspired other churches to follow, first through European ecumenical bodies, then through the World Council of Churches, and eventually through the Roman Catholic Church. It stresses that Creation Day honours God as Creator and commemorates the mystery of creation as a foundational mystery of Christian faith.

## Theological background for the Festival

The theological objectives of a liturgical Festival of Creation in Christ can be set out in several connected gestures.

First, the feast commemorates God's act of creation. The ecumenical Theological Overview speaks of creation out of nothing, continuous creation, and creation out of love. The world is not made from material which stands alongside God; it comes into being through God's free act. Nor is creation a remote event sealed in the past. God continuously sustains what he has made. Creation also flows from love: God creates from no deficiency, compulsion or need, but from the abundance of divine goodness.

Secondly, the feast is Christological. The Christian calendar unfolds the mystery of Christ, and the Festival of Creation in Christ brings into focus a dimension often present but seldom brought to the fore: the eternal Word as agent, meaning and end of creation. John 1, Colossians 1 and Hebrews 1 are central here. The Word is in the beginning; all things come into being through him; all things are created in him, through him and for him; he sustains the universe and holds it in being. The ecumenical Overview describes creation in Christ and new creation in Christ as two aspects of a single Christological arc. The one through whom all things were made is the same one through whom all things are redeemed and consummated. This is perhaps the most important interpretative key for the Festival: it celebrates the world's beginning in Christ in the light of its promised fulfilment in Christ.

Thirdly, the feast is Trinitarian. The Creed presents creation as the inseparable work of Father, Son and Holy Spirit. The Father is maker of heaven and earth; the Son is the one through whom all things were made; the Spirit is the giver of life. The Theological Overview emphasises that the feast offers the Church an opportunity to praise the Triune God as Creator, and to rebalance Christian imagination so that the Trinity is confessed in relation to creation as well as redemption. The new creation, too, is Trinitarian: promised by the Father, accomplished in the risen Son, and anticipated in the Spirit who is first-fruits.

Fourthly, the feast receives the created order as sacramental and covenanted gift. The Theological Overview describes creation as a sacramental reality which reveals and mediates the presence of its Creator. Matter and time are the basic grammar of Christian worship. The feast therefore makes explicit something on which all liturgy already depends. It also names creation as loved and covenanted by God, recalling the covenant with every living creature and the divine love which culminates in the Creator, in the incarnation, becoming a creature. In the sacraments, creation is already being gathered into the new creation: water becomes the sign of new birth; bread and wine are made a participation in the body and blood of the risen Lord.

*Hope for All Creation* deepens this account by giving it an Anglican ecclesial and pastoral frame. Drawing especially on the Psalms and Wisdom literature, the report describes the Earth as an interconnected creation sustained by God. Human beings belong within that creaturely fellowship. They are part of creation's interdependence, made of the same generic materials as other living things, dependent upon the same divine provision. At the same time, the document is careful about the distinctive vocation of humanity. Human beings are made in the image and likeness of God, but this vocation is not permission for exploitative dominion. It is, rather, a call to responsibility, care, and service. The language of 'rule' in Genesis is read in the context of God's own care for life and provision for the flourishing of creatures. The qualities demanded are those of a good shepherd rather than an owner who exploits. The Earth and its abundant life belong to the Maker, even when human beings are entrusted with responsibility for it.

This human vocation has become urgent because the created order is wounded by human sin. *Hope for All Creation* describes ecological crisis as a rupture in our relationships with God, with nature and with one another. It sees humanity's rejection of responsibility and defiance of creaturely place as disrupting the integrity of creation. Yet it also places this rupture within the larger hope of God's action in Christ: God reconciles all creatures to himself, promises the renewal of creation, and gives the Church freedom and responsibility in the Spirit to act in solidarity with the poor and the marginalised. The Church is called to active hope grounded in the resurrection.

The Paschal Mystery is therefore essential to the feast's theological horizon. As *Hope for All Creation* observes, Christ's blood reconciles all things, whether on earth or in heaven (Colossians 1.19–20). The cross is God's answer to the groaning of creation; the resurrection validates the material world and anticipates its transformation. Christ rises in a bodily manner as the first-fruits of what God intends for flesh, matter and creaturely life. God is renewing the world from within, with the risen Christ as first-fruits. The Festival of Creation in Christ should therefore be preached as a feast of both creation and reconciliation, and also as a feast of hope: the created world is destined for transfiguration, and the risen Christ is, as all Christians know, the beginning of that future.

## The liturgical texts

The Church of England has assigned the Festival of Creation in Christ to the first Sunday in September in order to give it prominence in regular Sunday worship, and to allow it to resonate with the understanding of Sunday as the day of resurrection. The date remains anchored to 1 September, thereby honouring the Eastern origin of the observance. Like other Festivals in the *Common Worship* Calendar, it may be transferred as long as it is kept.

### *Prayers of Penitence: Introduction to Confession and Kyrie Confession*

Echoing Psalm 24, this Invitation encourages us to think of the world as belonging to God before it is entrusted to us. We are invited to confess those times when we have treated this gift as something to be exploited and spoilt.

The Kyrie Confession develops this penitential gesture in a threefold form: God laid the foundations of the earth, while we have claimed the world as our own; all things were created through Christ and for Christ (Colossians 1.16), while we have not lived as those who belong to him; God makes all things new (Revelation 21.5), while our hope has faded. If all things are created through Christ and for Christ, then the misuse of creation is a failure to live under Christ's sovereignty. The confession ends by aligning repentance with the promise of new creation.

### *Collects*

Two Collects are provided. The first of these gathers together the whole theological premise of the Assisi process. It is Trinitarian in structure, Christological in emphasis, and doxological in purpose. The prayer asks that worshippers may be drawn into the mystery of divine love and, with all creation, proclaim God's praise. The Church is joining the praise which creation already offers by existing before God, and anticipating the praise of the renewed creation in which all things are gathered into Christ.

The alternative collect gives the same theology a more immediately pastoral idiom, and was drafted with a particular attention to accessible language.

### *Lectionary*

The lectionary provision gives the Festival a rich scriptural reservoir. Psalm 19, Genesis 1.1–5, 26–31, and Hebrews 1.1–4 are appointed for first Evening Prayer. Genesis supplies part of the classic narrative of the creation of the heavens and the earth, with Hebrews reminding us of the presence of Christ in creation. Psalm 19 invites even the heavens declare the glory of God. The Principal Service then appoints Isaiah 55.12–end, Psalm 96, Colossians 1.15–20 and John 1.1–14, with Matthew 8.23–27 as an alternative Gospel. This is a carefully constructed sequence. Isaiah imagines mountains, hills and trees drawn into joy; Psalm 96 summons the whole earth to worship; Colossians presents Christ as the image of the invisible God, the firstborn of all creation, and the one in whom all things hold together; John proclaims the Word through whom all things came into being and who became flesh. The appointed readings press towards Christ as the one in whom creation coheres and through whom peace is made.

The remaining readings make the theme of new creation still clearer. Ezekiel 47 offers the river flowing from the temple; Revelation 22 speaks of the tree of life and the healing of the nations; John 15 presents Christ as the true vine. Job 38 places humanity before the majesty and mystery of God's creative wisdom; Sirach offers the splendour of divine majesty and the glory of the heavens; Psalm 148 summons the whole creation to praise; Revelation 4 shows the heavenly host worshipping the Creator; 2 Peter points towards a new creation where righteousness dwells. The effect is catechetically powerful. Creation is shown to be a theme across the whole scriptural canon: beginning, wisdom, prophecy, Gospel, apostolic witness and apocalypse. This will help preachers avoid reducing the feast to Genesis, or the prologue to John, alone.

### *Gospel Acclamation*

The Gospel acclamation comes from Revelation: 'You are worthy, our Lord and God, to receive glory and honour and power, for you created all things.' Placed immediately before the Gospel, it offers a foretaste of heavenly worship.

(While the provision for Christ in Creation does not include any new form of Intercession, a range of suitable forms can be found on pages 65 to 71 of this volume.)

### *Introduction to the Peace*

The invitation to share the peace of Christ is framed by Isaiah 55: the people shall go out in joy and be led back in peace, while mountains and hills burst into song and the trees clap their hands. The exchange of peace is a sign of that larger reconciliation into which God draws the whole creation.

### *Prayers at the Preparation of the Table*

The first prayer blesses God who filled the world at its creation with plenty, and asks that the gifts of bread and wine might teach gratitude and faithful stewardship. This is exactly the sacramental vision identified by the theological rationale for the Festival; in other words, that created gifts become the means by which the Church learns to receive the world rightly.

The alternative prayer from the *Didache*, which has in an adapted form been a staple of *Common Worship*, turns the table into an image of the Church gathered from the corners of the earth into the kingdom. The bread and wine belong to the first creation, yet in the Eucharist they become signs and instruments of the new creation already breaking in.

## *Prefaces*

Perhaps the clearest eschatological vision of the Festival is provided in these Prefaces. In the first Preface, in Christ the risen Lord 'the new creation is being brought to perfection; a broken world is being renewed; and creation itself will share in the glorious liberty of the children of God.' This text is especially important in the light of *Hope for All Creation*: it answers grief and anxiety about the damaged earth and its fruits with hope in Christ. This preface offers a framework for understanding the Festival: the Church celebrates the Creator by looking to the risen Christ in whom the new creation is already being brought to perfection.

The Extended Preface is the most complete eucharistic statement of the Festival's theology. It declares that the heavens are made by God's word and their host by the breath of his mouth (Psalm 33). It names humanity as formed in God's image (Genesis 1.26–28) and entrusted with the works of God's hands (Psalm 8.5–8). It then turns to the Son, the firstborn of all creation (Colossians 1.15), who holds all things together (1.17) and through whom God makes all things new (Revelation 21.5). Finally, it invokes the first-fruits of the Spirit as creation waits with eager longing for the glorious freedom of the children of God (Romans 8.19–23).

This preface deserves particular attention in preaching and teaching. It contains, in compressed but scripturally inspired form, the whole doctrine which the feast seeks to teach. Its structure is not merely concerned with origins or fruits of creation; it is also eschatological, and concerned with fulfilment. The prayer moves naturally into the Sanctus, where the congregation, together with the whole created order, joins with the ceaseless worship of heaven.

## *Post Communion*

This prayer, drawn from existing *Common Worship* provision, brings together the tree of life and the bread of life. God the Creator set the tree of life at the heart of earthly paradise and, equally, the bread of life at the heart of the Church; those nourished at the earthly table ask to be transformed by the glory of the Saviour's cross and to enjoy the delights of eternity. The tree of life evokes creation's first gift and lost communion; the bread of life evokes Christ's eucharistic gift in the Church; the Cross is named as glory. The material world is therefore read through sacrament and redemption. The prayer also carries the whole congregation from the earthly paradise to the holy table, and from the holy table to the delights of the new creation.

### *Act of Commitment*

Appearing first in the original edition of *A Time for Creation*, this offers the congregation an opportunity to make a formal promise within worship to engage in what we are all sent out to do.

### *Blessing*

The Blessing sends the congregation out in delight and wonder under the sign of the Trinity. Its reference to risen life is especially significant: the congregation is sent not merely to preserve what remains, but to witness to the renewing life of Christ.

### *Proper texts for Daily Prayer*

The Daily Prayer provision completes the picture. Morning Prayer may use *A Song of the New Creation*, and Evening Prayer may use *A Song of Francis of Assisi* (aptly for a Festival whose origins owe much to meetings in Assisi). The antiphons for the Gospel canticles gather the main themes once again: the heavens made by God's word and their host by the breath of his mouth; the Spirit of God making us and the breath of the Almighty giving life; and the heavenly acclamation of God's worthiness because he created all things. The appointment of *A Song of the New Creation* is particularly apt, because it ensures that Morning and Evening Prayer for this Festival is not confined to the first creation, but to convey the promise that God is making all things new.

* * *

The Festival of Creation in Christ should, it will now be clear, be celebrated as a gift above all else, for its first movement is adoration. It invites the Church to praise the Triune God for creation; to behold Christ as the Word through whom all things were made; and to receive the Spirit as giver of life. In this light, we confess the sin by which we have misused God's gift; at Holy Communion we receive bread and wine as signs of creation offered back to God and returned as communion; and to go out in the hope that the risen Christ is renewing a broken world.

*Hope for All Creation* is clear that care for creation belongs within Christian mission. It connects the protection of God's world with, in the words of the Five Marks of Mission, proclamation, teaching, response to human need, transformation of unjust structures, and peace and reconciliation. It also stresses that the sacraments are embodied signs of grace working through material creation, and that the Church's life of Word and sacrament nourishes care for creation as an integral dimension of discipleship.

## Holy Cross Day (14 September)[10]

### Festival – Red

The cross on which our Lord was crucified has become the universal symbol for Christianity. After the end of the persecution era, early in the fourth century, pilgrims began to travel to Jerusalem to visit and pray at the places associated with the life of Jesus. Helena, the mother of the emperor, was a Christian and, whilst overseeing excavations in the city, is said to have uncovered a cross, which many believed to be the Cross of Christ. A basilica was built on the site of the Holy Sepulchre and dedicated on this day in the year 335.

On this feast day, set as it is in September and potentially therefore in the middle of a creation season, there is an opportunity to reflect on the cosmic significance of the cross. In Christian liturgy, the cross is often represented as the tree of life, a sign of the new life that comes to us through Christ's death and resurrection. In preparing the disciples for the fate that awaited him, Jesus spoke of his impending suffering and death on the cross as being like a grain of wheat that falls into the earth and dies: 'if it dies, it bears much fruit.' (John 12.24).

*Collect*

Almighty God,
who in the passion of your blessed Son
made an instrument of painful death
to be for us the means of life and peace:
grant us so to glory in the cross of Christ
that we may gladly suffer for his sake;
who is alive and reigns with you,
in the unity of the Holy Spirit,
one God, now and for ever.

*All* **Amen.**

*Readings*

| | |
|---|---|
| **Old Testament** | Numbers 21.4–9 |
| **Psalm** | 22.23–28 |
| **New Testament** | Philippians 2.6–11 |
| **Gospel** | John 3.13–17 |

[10] For full liturgical provision, see *Common Worship: Festivals* pp.102–107

## *Post Communion*

Faithful God,
whose Son bore our sins in his body on the tree
and gave us this sacrament to show forth his death until he comes:
give us grace to glory in the cross of our Lord Jesus Christ,
for he is our salvation, our life and our hope,
who reigns as Lord, now and for ever.

***All*** **Amen.**

## *Short Reflections*

*Hymns on the Cross by Venantius Fortunatus*

O Faithful Cross, amongst all trees the only noble Tree,
No other wood has grown a tree of such leaves, flowers, and fruit.
O sweet wood, sweet nails, how sweet the burden that you bear.

*Crux Fidelis*

O sweet and noble wood,
on your boughs you bear fruit so rare.
You are planted where the water flows,
and your spreading branches are adorned with flowers.
On your branches hangs the vine
from which blood-red sweet wine flows.

*Crux Benedicta*

*'In Adoration of the Cross' by Saint Theodore of Studios*

How precious is the gift of the cross, how splendid to contemplate! In the cross, there is no mingling of good and evil, as in the tree of paradise: it is wholly beautiful to behold and good to taste. The fruit of this tree is not death but life, not darkness but light. This tree does not cast us out of paradise, but opens the way for our return. This was the tree on which Christ, like a king on a chariot, destroyed the devil, the lord of death, and freed the human race from tyranny. This was the tree upon which the Lord, like a brave warrior wounded in hands, feet and side, healed the wounds of sin that the evil serpent had inflicted on our nature. A tree once caused our death, but now a tree brings us life.

*A Meditation on the Holy Cross by Saint Anselm*

Holy Cross, recalling the cross of our Lord Jesus Christ,
that brings us back from the abyss of eternal death
and leads us to the eternal life we had lost by sin.

I glory in that cross, and by the cross I adore our merciful Lord
and what in mercy he has done for us.

The cross by which we are sealed for God,
glorious cross, we ought to glory only in you.
By you the world is renewed and made beautiful with truth.

*Abridged*

## Propers for St Francis (4 October)

**Lesser Festival – Religious – White**

*Collect*

O God, you ever delight to reveal yourself
to the childlike and lowly of heart:
grant that, following the example of the blessed Francis,
we may count the wisdom of this world as foolishness
and know only Jesus Christ and him crucified,
who is alive and reigns with you,
in the unity of the Holy Spirit,
one God, now and for ever.

*All* **Amen.**

*Old Testament* Micah 6.6–8

*Responsorial Psalm*

*R* **Be joyful in the Lord, all the earth:**
**[give thanks and bless his name].**

O be joyful in the Lord, all the earth;
serve the Lord with gladness
and come before his presence with a song. ***R***

Know that Lord is God;
it is he that has made us and we are his;
we are his people and the sheep of his pasture. ***R***

Enter his gates with thanksgiving
and his courts with praise;
give thanks to him and bless his name. ***R***

For the Lord is gracious;
his steadfast love is everlasting,
and his faithfulness endures from generation to generation. ***R***

*Psalm 100*

*New Testament* Galatians 6.14–18

*Gospel* Luke 12.22–34

*Post Communion*

Lord God,
you made your church rich
through the poverty of blessed Francis:
help us like him to sing creation's praise,
and care for all your creatures;
through Jesus Christ, our crucified and risen Lord.

*All* **Amen.**

# OTHER RESOURCES

## Acclamations

1 Where were you when I laid the foundations of the earth
and established its boundaries?
**Lord, your purpose stands for ever.**

Where were you when the morning stars sang together
and all the heavenly beings shouted for joy?
**Lord, your purpose stands for ever.**

Who shut up the sea behind doors
when it burst forth from the womb,
when I made the clouds its garment?
**Lord, your purpose stands for ever.**

Have you commanded the morning
and caused the dawn to know its place?
**Lord, your purpose stands for ever.**

From whose womb does the ice come forth
and who gives birth to the hoar-frost of heaven?
**Lord, your purpose stands for ever.**

Who has endowed the heart with wisdom,
and given understanding to the mind?
**Lord, your purpose stands for ever.**

2 God be gracious to us and bless us;
and make his face to shine upon us.

That your way may be known upon earth,
your saving power among all nations.

**Let the peoples praise you, O God;**
**let all the peoples praise you.**

O let the nations rejoice and be glad,
for you will judge the peoples righteously
  and govern the nations upon earth.

**Let the peoples praise you, O God;**
**let all the peoples praise you.**

Then shall the earth bring forth her increase,
and God, our own God, will bless us.
God will bless us,
and all the ends of the earth shall fear him.

**Let the peoples praise you, O God;**
**let all the peoples praise you.**

3 **The earth has yielded its harvest;**
**God, our God, has blessed us.**

You visit the earth and water it;
**you make it very plenteous.**

You soften the ground with showers;
**and bless the increase of it.**

You crown the year with your goodness;
**and your paths overflow with plenty.**

The meadows are clothed with sheep;
**the valleys stand so thick with corn,**
**they shout for joy and sing.**

**The earth has yielded its harvest;**
**God, our God, has blessed us.**

## Psalm Prayers

8 We bless you, master of the heavens,
for the wonderful order which enfolds this world;
grant that your whole creation
may find fulfilment in the Son of Man,
Jesus Christ our Saviour. Amen.

65 May the richness of your creation, Lord,
and the mystery of your providence
lead us to that heavenly city
where all peoples will bring their wealth,
forsake their sins and find their true joy,
Jesus Christ our Lord. Amen.

96 Lord God, you draw us by your beauty
and transform us by your holiness;
let our worship echo all creation's praise
and declare your glory to the nations;
through Jesus Christ our Lord. Amen.

98 Lord God, just and true,
you make your salvation known in the sight of the nations;
tune the song of our hearts to the music of creation
as you come among us to judge the earth;
through our Saviour Jesus Christ. Amen.

126 Lord, as you send rain and flowers
even to the wilderness,
renew us by your Holy Spirit,
help us to sow good seed in time of adversity
and to live to rejoice in your good harvest of all creation;
through Jesus Christ our Lord. Amen.

## Readings on Creation and the Environment

### Old Testament Readings

| | |
|---|---|
| A cosmic hymn of praise | Genesis 1.1–2.3 |
| The garden of creation | Genesis 24b–19 |
| Creation restored after the deluge | Genesis 8.12–22 |
| God's promise to sustain creation | Genesis 9.8–17 |
| God's provision | Exodus 16.11–15 |
| A feast of fruitfulness | Leviticus 23.33–43 |
| The promise of a fertile land | Leviticus 26.1–13 |
| God's gifts to the redeemed | Deuteronomy 6.9–14 |
| The gift of the land | Deuteronomy 8.7–18 |
| Offering of the first-fruits | Deuteronomy 26.1–11 |
| The promise of restoration | Amos 9.13 |
| The covenant of peace with all creation | Hosea 2.18–23 |
| Repentance and restoration | Hosea 14.2–10 |
| The harmony of animal kingdom in the messianic age | Isaiah 11.1–9a |
| Feasting in God's presence | Isaiah 25.6–9; 55.6–11 |
| The blossoming of the wilderness at God's appearing | Isaiah 35.1–7 |
| The whole cosmic scope of praise for God's redemption | Isaiah 55.12–end |
| A sign of a fruitful future | Jeremiah 32.6–15 |
| God promises his people a peaceable land | Ezekiel 34.25–31 |
| The hope of humankind restored | Ezekiel 37.1–14 |
| A vision of waters restoring the land | Ezekiel 47.[1–6a] 6b–12 |
| The wonder of creation | Ecclesiasticus 42.15—43.12 |
| A garden as symbol of divine love | Song of Songs 4.12–13 |
| Kinship and the land | Ruth 2.1–13 |
| Defiant nature and the hope of immortality | Job 14.7–10 |
| The rich minerals of the earth | Job 28.1–6 |
| God's majesty and creation's marvel | Job 38.1–11, 16–18 |
| God's wisdom and creation | Proverbs 8.1–4, 22–31 |

## Readings from the Apocrypha

| | |
|---|---|
| The wonders of creation | Ecclesiasticus 42.15–33 |
| Blessing of the forest | Judith 16.13–15 |
| The praises of creation | Song of the Three, vv.28–59; Jubilees 2.21–27 |

## New Testament Readings

| | |
|---|---|
| The natural world longing for liberation | Romans 8.18–22 |
| The risen Christ appears in the garden, prefiguring a new creation | John 20.15 |
| Christ as the new creation | Colossians 1.1–15 |
| Creation renewed for the healing of the nations | Revelation 22.1–5 |
| God's generosity and human gratitude | 2 Corinthians 9.9–11a |
| Sufficiency of resources and simplicity of lifestyle | 1 Timothy 6.7–10, 17–19; *or* Matthew 6.25–34 |
| The risen Christ and the abundance of life | John 21.1–14 |
| The promise of paradise restored | Revelation 2.1–7 |

## Gospel Readings

| | |
|---|---|
| The book of nature | Matthew 6.25–34 |
| The Lord of nature brings peace to violent discord | Mark 4.35–41 |
| Nurturing the fig tree | Luke 13.6–9 |
| From the seed to a tree of life: an image of God's reign | Mark 4.30–32 (*or* Luke 13.18–19; Matthew 13.31–32) |
| Hungry people fed on Christ's abundance | John 6.1–13 |
| God's people as a fruitful vine | John 15.1–17 |

# APPENDIX – OTHER SOURCES OF MATERIAL

*Church of England website Creationtide Resources* (includes discussion materials and other suggestions) https://www.churchofengland.org/more/policy-and-thinking/our-views/environment-and-climate-change/creationtide/creationtide

*Church Support Hub*: https://churchsupporthub.org/?s=Harvest

*Celebrating Festivals Together*: https://spckpublishing.co.uk/festivals-together

*Creative Ideas for Wild Church: Taking all-age worship and learning outdoors* Mary Jackson & Juno Hollyhock, Canterbury Press 2016: https://chbookshop.hymnsam.co.uk/books/9781848258815/creative-ideas-for-wild-church

*God's Good Earth: Praise and Prayer for Creation* by Anne & Jeffery Rowthorn, Liturgical Press 2018: https://chbookshop.hymnsam.co.uk/books/9780814644126/gods-good-earth

*Hope for All Creation: A theological response to the environmental crises*, The House of Bishops of the Church of England, Church House Publishing 2026. Supporting digital resources available via cofe.io/HopeForAllCreation

*Together for a Season: All-age resources for Feasts and Festivals of the Christian Year* Ed. Gill Ambrose, CHP 2009: www.chpublishing.co.uk/books/9780715140642/together-for-a-season-feasts-and-festivals

The website of the *Anglican Communion Environmental Network* (ACEN) gives links to documents and resources from a number of areas of the Anglican Communion. See http://acen.anglicancommunion.org/resources.documents.cfm

The *Anglican Church of Southern Africa* has provided a substantial collection of study and reference material and resources for worship. This can be found on the ACEN website at: http://www.anglicancommunion.org/resources/docs/season–of–creation.pdf

The *Anglican Church of Australia* also has a fine collection of liturgical material on the ACEN website.

*The Episcopal Ecological Network* (TEC) offers online resources for the Season of Creation, normally observed during the Season of Pentecost. This has not been updated since 2011. See http://econline.org/reflect/liturgy/com

*CTBI* has resources entitled '*Time for Creation*' (1 September–4 October). See www.ctbi.org.uk

The *European Christian Environmental Network* provides material at: http://ecen.org/cms/index.php?page=creationtime

The Report by ChurchCare (The Cathedrals and Church Buildings Division), entitled *Shrinking the Footprint*, includes worship and study resources: http://www.churchcare.co.uk/shrinking–the–footprint

*Alliance of Religions and Conservation* – resources: http://www.arcworld.org/news.asp?pageID=408

*Sharing Eden* – a book setting out interfaith responses to environmental and creation–related issues: http://www.sharingeden.org/

*Operation Noah* provides a good variety of material for worship. Though this is not authorised by the Church of England, it comes recommended by Diocesan Environmental Officers: http://www.operationnoah.org/sites/default/files/Liturgy_materials_0.pdf

Dr Jeremy Clines, Chaplain of Sheffield University, has published his PhD thesis, Earthing Common Worship: an ecotheological critique of the Common Worship texts of the Church of England (Birmingham, 2011) at http://etheses.bham.ac.uk/2838/

www.ingramcontent.com/pod-product-compliance
Lightning Source LLC
LaVergne TN
LVHW100325210926
848480LV00003B/1261

* 9 7 8 0 7 1 5 1 2 4 4 2 0 *